Texts in Philosophy

Volume 34

Doing Logic
The Continuum through Sets

Volume 24
Karl Popper. A Centenary Assessment. Volume I. Life and Times, and Values in a World of Facts
Ian Jarvie, Karl Milford and David Miller, eds

Volume 25
Karl Popper. A Centenary Assessment. Volume II. Metaphysics and Epistemology
Ian Jarvie, Karl Milford and David Miller, eds

Volume 26
Karl Popper. A Centenary Assessment. Volume III Science
Ian Jarvie, Karl Milford and David Miller, eds

Volume 27
Unorthodox Analytic Philosophy
Guillermo E. Rosado Haddock

Volume 28
Quantum Heresies.
Kent A. Peacock, with a foreword by James Robert Brown

Volume 29
Carrollian Notes
George Englebretsen

Volume 30
Problems from Hume
David Bostock

Volume 31
The Labyrinth of Infinity or the Enigma of Existence
Stathis Livadas

Volume 32
Experience and the Ultimate Structure of Reality, Huserl's Pursuit of Truth
Claire Ortiz Hill

Volume 33
Analytic Philosophy: A User's Guide
Andrew David Irivine

Volume 34
Doing Logic. The Continuum through Sets
Montgomery Link

Doing Logic
The Continuum through Sets

Montgomery Link

ISBN 978-1-84890-509-2

College Publications
Scientific Director: Dov Gabbay
Managing Director: Jane Spurr

http://www.collegepublications.co.uk

Original cover design by Laraine Welch
Cover image: Porte de la Casbah by Henri Matisse.
Image obtained from Wikipedia

To My Parents

It may be we shall touch the Happy Isles,
And see the great Achilles whom we knew.
Tennyson, "Ulysses"

Thanks to Raffaele Pisano for comments on the frame of the argument, for edits, and for building the originals of both the index and the index of names. Thanks to Brian F. Kiniry for helpful conversations and a close reading of the initial proposal, with many suggestions for improvement, to Joseph X. Donovan, to my current students in Advanced Formal Logic, and to Trevor Stahl Link. Thanks to the editors of the Philosophy Series at College Publications and to Jane Spurr. Errors are mine.

Bullfinch Triangle and Beacon Hill, Boston
The Ides of March, 2026

Contents

CHAPTER 1

Introduction

What is continuity? Look at a continuous line, for example. Split it in two and now there are two continuous lines. Is it *lines all the way down*? But the line was cut at a particular point. Is the line, then, made up of indivisible points? The answers to these questions remain hotly debated.

My title suggests a close connection between logic and sets. Although this might seem counterintuitive to some, I am not claiming that sets are part of logic but rather, truth be told, that when doing logic, one is soon confronted with sets. Actually, when one is doing many kinds of research, one quickly runs into sets. Think about it, when one is doing research in the sciences, in thermodynamics or quantum field theory, let alone mathematics, one quickly runs into sets. And yet, traditionally, sets have been entia non grata, perhaps even more despised than relations.

That is why, when someone insists that the ordered pair is the paradigm of philosophical analysis, it can be regarded as a bit of a joke. Philosophers from Aristotle onward have seen the relation as a different category of being. Now, the analysis of the relation to the ordered pair is a first step to clearing things up, but let us take another step. We will see that the ordered pair is a set. When doing logic, one runs into sets. So, the title, *Doing Logic. The Continuum through Sets*, works, as the reader hopefully will come to agree.

The appropriateness of the title aside, we want to look at a particular special question, a foundational and philosophical question. Is the continuum real? We will see it explicated as the number line. These real numbers on the number line have definitions based on sets. So, are those sets real? Does the theory of sets deal with real or ideal objects?

That is not an idle question, actually. Continuity represents a juncture between mathematical and physical reality. For example, the idea of a function f being continuous at a is fundamental to the calculus.

The calculus is what enables us to give precise answers to questions about physical reality. Descartes made an incredible advance when he was able to analyze geometry. Think of our

Cartesian coordinate system, to capture his influence. And yet, he was able to model only static objects, so he was not able to capture real motion. The calculus had to be introduced before that could happen. Continuity in one respect is captured in the calculus.

What makes the continuum? Is it part of physical reality? Or is it ideal and as such part of the constructed world? A closely connected issue is that of infinity. I am a finite being. How does the continuum appear to me, a finite being? The continuum might appear finitistic, even if not entirely finite, but on the usual analysis, it is actually infinite. Another closely related issue is divisibility. If the continuum is real, and that itself is highly debated, are the points that compose its reality, on one reduction, real? What about the set axioms that provide for the continuum? Are sets a constitutive part of foundational reality? We look at the continuum from our finite perspective and answer that we are open-minded on sets.

After all, in this day and age, if one is willing to accept the physical reality of photons, for example, and of Compton Scattering, for example, perhaps it is not so hard to conceive of the mathematical reality of sets, for example, and the continuum, for example. We might draw a line in the sand, if that helps, and talk only about a highly constructed universe of sets, sometimes called L. So that by the reality of the universe of sets one commits to something like truths in this model. The trouble with calling this a reality is that there is another kind of model, a forcing model, that has no more axiomatic requirements than L and yet it gives quite different results about the continuum. Perhaps a new model of the universe is required. Maybe, for example, by extending L.

One avenue to deductive thought is through the logic of symbols. The logic of symbols is a structure for deductive thought. We casually introduce the canonical language but without going too much into the axiomatics. Once the continuum is satisfactorily defined, further investigations are then based on this unity, introducing a philosophical topic, which we lurch into toward the end of chapter five. This topic is the nature of the underlying reality of the natural world.

We can consider another example from quantum theory. It is quantum reality and the interplay of electrons that determines the elements chemists utilize. Yet, to understand water, the chemist would not necessarily need to return to physics for the explanation of the make up of hydrogen and oxygen. If we have as a principle that always the goal is to identify the underlying reality, do we mean by that the elements? Likewise with the continuum and the basement reality of set theory, one wants to ask what exactly one is getting into by believing in reality. Is it the surface reduction, or do one's commitments run deeper? The analysis gives logical clarity, but the canonical notation necessary for the continuum analysis includes not only logic but also sets.

Accordingly, chapter two looks at some of the basic concepts of sets. We focus on membership, inclusion, and extensionality. We introduce the standard set builder notation, also known as set abstraction. In chapter three we expand on the basics. We describe equinumerosity and some fundamental set operations. We introduce the counter sets. Then we turn to the foundations and the hierarchy of sets.

In chapters four and five we extend the foundational investigation to the continuum. I once covered much of this material in an essay entitled "The Dispute between Two Accounts of the Continuum."* Here I argue that the continuum has a unity more than is captured in a conglomerate concept of the continuum, something not only ideal but also real that is unlocked in the analysis.

What about these theoretical constructs we have introduced to investigate the continuum? What is their standing? How am I to understand their reality? Chapter six focuses on the approach to reality from naturalism, while chapter seven does the approach from linguistic analysis. The penultimate chapter ties up a loose end.

Overall, the canonical notation reveals our ontological commitments. It appears to be telling us something about what we believe. We seem not to want to admit the commitments, we have scruples about this, after all; however, it appears to be saying that sets are part of the reality of the continuum. Thus, insofar as the continuum is not just a mathematical reality but also a physical reality, sets are part of physical reality.

---

* [**66**].

CHAPTER 2

Set

1. Theory

1.1. Sets are the building blocks for the formal representation of the grounds or foundations of mathematical reality. Sets are also used for the mathematical representation of physical reality and the foundations of quantum theory.

1.2. If sets are real, they are often puzzling. For example, there is an ancient aporia about a philosopher and his dun horse on the road. The ancient question was, "How many are there?" The *paradoxical* answer is that there are three, counting the philosopher, his dun horse, and the set containing the pair of them.

The seeming paradox is clarified once one recognizes that sets can be counted, like any other object. There is a general truth about sets, that they quickly outnumber the number of objects given. In our case above, we have the pair set containing both the philosopher and the horse, but there is also another set, a one-membered set, that contains just the dun horse. We call a one-membered set a *singleton*. Already we are counting and working with sets.

1.3. What is *set theory*? Let's check with an expert. In 1908, Ernst Zermelo gave the first official axiomatization of the theory of sets in his magnificent "Investigations in the Foundations of Set Theory I." Here is the first sentence of it:

> Set theory is that branch of mathematics whose task is to investigate mathematically the fundamental notions of "number", "order", and "function", taking them in their pristine, simple form, and to develop thereby the logical foundations of all of arithmetic and analysis.*

In this account, I will present some *gentle* work with sets, the only idea being to toy around with sets and gain some familiarity with them. The focus is on understanding the machinery of sets insofar as that will facilitate the logic of symbols. Strictly, we do

* Trans. Bauer-Mengelberg in [**117**], p. 200.

not work in set theory. Instead, we are using the apparatus to build up our symbolic resources.

2. Membership

2.1. The theory of sets takes membership as primitive and symbolizes it ‘ $\in$.’ In set theory, sets are defined by their members. The axioms of set theory are stated using the membership predicate. For example, one axiom of what is called Zermelo-Fraenkel set theory (ZF) asserts the existence of the empty set, and that can be stated as the formula there exists a set X that has no members: if A is a set, then A is not in X. Here we see an assertion about existence.

2.2. Let me be a bit more precise with the extensional viewpoint. Membership is a two-place predicate, whereby x is a member of y, or $x \in y$ for two sets x, y. Sometimes one wants to talk about a collection that is not really a set, because it exceeds the scope of what the axioms can generate, but it is still very clear, "clear" in a precise sense, and can be whittled down to a set with no worries that something key gets left out, or something extraneous gets added in.

Let us call those collections *classes.* For example, consider the idea of gathering together all of the sets in the universe of sets.

That collection is not a set, not really. Were it a set, it would contain itself as a member. But that would lead to an infinite downward spiral of sets within sets. This outcome is barred by what is known as *regularity* or the Axiom of Foundation, which states that every nonempty set X has a member that does not intersect with X. As such, there are no infinite downward chains of membership. Membership is basic.

In cases of extreme membership, where a collection has so many members that we do not seem to have an axiomatic handle on it, we call the collection a *class* instead of a set. Classes remain outside the realm of sets without further specification. Let us come back to this later. The usage is specialized for the work we are doing. In ordinary English, it is okay to use either term to talk about a formal collection.

2.3. Even without entering into set theory itself, and while remaining in some other ambient theory system, we can utilize the canonical language of first order logic in a casual way to speak about sets and classes of things casually and without yet committing to that entire reality.* We use a set building operation to talk about simulation set. This is not the difference between sets and classes that I mentioned above. That was like replacing one

---

* [**86**], p. 4.

sort of thing, sets, with another sort of thing, classes, when we do not have sufficient purchase on the collection to be sure that it is sharply defined.

Set abstraction is a remarkable tool. The simulation is accomplished by the notational device of set building. The set building operation saves the appearances, making a quantification over sets or classes seem to be about not sets but ersatz simulations, so that the conversation, if not about sets, is not about anything at all.

Why bother with this question? Why worry about the reality of sets? Well, having been teaching this material for some years, I am well aware of the skeptical mindset of many undergraduates. Many students will not want me to try to convince them that, to go ahead with a course using this book, they must commit to the reality of sets. Instead, they will want to know that they can learn the stuff and remain agnostic on the philosophical questions until some other time. The idea of the simulation is to show how one can delay ontological concerns without sacrificing clarity and accuracy. But strictly speaking one when doing set theory must prove that there is a set that exists such that each of its members fulfills the formula of abstraction. When I was growing up, they were teaching the "new math" to young people. They do not do that anymore, or not as often. We might review some of the basic concepts and operations that are fundamental to this approach to mathematics, whether they teach it anymore or not.

3. Extensional Identity

3.1. As we have already seen, ordinarily we do not make much of a difference between a set of things and a class of things. Both are ways to talk about a collection. The difference arises only in the infinite case when the collection becomes greater than a set can be. So, in what follows we will stick with the term 'set' for a formal collection. What a set *collects* are its elements. A set is determined by its elements. The elements or members are what distinguish one set from another.

Consider the singleton that has one member, the object o. A singleton, by the way, is a set with only one element, a pair is a set with two elements, and so on. For our investigation of deductive reasoning, we need to be able to handle the notion that o is a member of $\{o\}$, for which we will write $o \in \{o\}$.* Assume $X = \{o\}$. Since X is a singleton having only one member o, X is not an element of X, which we will write $X \notin X$.

As I have said, there is an axiom, the Axiom of Foundation, that rules out self-membership for any set. Membership is not a reflexive relation. Nor is it symmetric. The United States is a

* [**33**].

member of the United Nations but the United Nations is not an element of the United States. Further, it is not transitive:

$$x \in \{x, y\},$$

and

$$\{x, y\} \in \{\{x, y\}, \{y, z\}\},$$

but

$$x \notin \{\{x, y\}, \{y, z\}\}.$$

3.2. Sets can take any objects as members. For example, compare the set X of the first five presidents of the United States with the set Y of the American Founding Fathers who served as president of the United States. Not only are these sets the same *size*, by which I mean that they have the same cardinality or number of members, they are *equal*: They have exactly the same members, $X = Y$. Witness:

$$\{\text{Washington, Adams, Jefferson, Madison, Monroe}\}.$$

Let us establish a principle:

EXTENSIONALITY 1. *$X = Y$, if they have the same members.*

We can rephrase our first principle in a more symbolic fashion, once we clarify that in what follows, the expression 'iff' is not a typo. That expression stands in for the logical idea of 'if and only if.' Now we specify that for two sets X and Y and every object o:

EXTENSIONALITY 2. *Suppose*

$$o \in X \qquad \text{iff} \qquad o \in Y$$

Then $X = Y$.

For example, the set containing the natural numbers between 1 and 4 and the set containing the first two prime numbers are the same.

3.3. We can quickly see that this definition captures the three key features of the identity relation, (i) reflexivity, (ii) symmetry, and (iii) transitivity. (i) For every set X, X has the same members as itself, so

$$X = X.$$

(ii) For every set X and Y, if X has the same members as Y, then Y has the same members as X, so

$$X = Y \qquad \Rightarrow \qquad Y = X.$$

Here the $\Rightarrow$ symbol is introduced to represent a conditional connection between antecedent and consequent: If $X = Y$, then $Y = X$. (iii) For every set X, Y, and Z,

$$(X = Y \,\&\, Y = Z) \quad \Rightarrow \quad X = Z.$$

Again, what an expression of equality tells us is that here we have two expressions for the same object:

> Incidentally, we write "$A = B$" to mean that A and B are the same object. That is, the expression "A" on the left of the equality symbol names the same object as does the expression "B" on the right. If $A = B$, then automatically (i.e., by logic) anything that is true of the object A is also true of the object B (it being the same object). For example, if $A = B$, then it is automatically true that for any object t, $t \in A$ iff $t \in B$. (This is the converse of the principle of extensionality.) As usual, we write "$A \neq B$" to mean that it is not true that $A = B$.*

3.4. Let me introduce the helpful semicolon operation. This operation joins to a set X a new member o:

$$X; o \;=\; X \cup \{o\}.$$

The members of the new set are o and the members of X. If $o \in X$, then the operation leaves everything as it is, nothing is adjoined to X:

$$X; o \;=\; X \qquad \text{iff} \qquad o \in X.$$

Elementhood is not repetitive, for, by Extensionality, $\{a, a, b, c\} = \{a, b, c\}$. There is more on the union operation below.

4. Inclusion

4.1. One can think of a set with no members, for example the set of all frogs that are sheep, the set of all presidents of the United States that served at the Battle of Bunker Hill, or the set of all the natural numbers that are less than zero. Call it the *empty set,* $\emptyset$, or the *null set.* It is also symbolized as Λ. It is established by an axiom. Besides the fact that (α) no set is a member of the empty set, some other facts about the empty set are (β) the empty set is a member of some sets and is not a member of others; (γ) the empty set is a subset of every set; however, (δ) the only set that is a subset of the empty set is the empty set.

* [**33**], p. 2.

We mentioned that a singleton is a one-membered set $\{x\}$. If $x = \emptyset$, then its successor is $\{\emptyset\}$. "Okay, $0, 1, \ldots; \emptyset, \{\emptyset\}, \ldots$. What about 2?"

There is a pair set $\{0, 1\}$ that contains just 0 and 1. A *pair set* is a set $\{a, b\}$ such that

$$x \in \{a, b\} \qquad \text{iff} \qquad x = a \text{ or } x = b.$$

But $\{0, 1\} = \{1, 0\}$, and in general $\{x, y\} = \{y, x\}$, i.e., the pair set is *unordered.*

The pair set is not as innocuous as it might at first blush appear. Once combined with only one other axiom, that of Replacement, this pair of axioms is adequate to support all the other eight axioms of set theory. What is Replacement? Only the idea that one can replace a class with the image of a function, if that function is very carefully defined, the image being a genuine set.

Later below we will look at the ordered pair $\langle 0, 1\rangle$. Let's come back to the natural numbers later.

4.2. When I cited some facts about the empty set a few paragraphs ago, I mentioned the subset relation. The set of the first five presidents of the United States is a subset of the set of persons born before the invention of the steam engine. Again, the set of the first five United States presidents is not included in the set of those who signed the Constitution of the United States of America.

In general, for any two sets X and Y and all objects o,

INCLUSION 1. *Suppose*

$$o \in X \qquad \Rightarrow \qquad o \in Y$$

Then $X \subseteq Y$.

So, for example

$$\{a, b\} \subseteq \{a, b, c\},$$

and as we will see below, the integers include the natural numbers:

$$\mathbb{N} \subseteq \mathbb{Z}.$$

We can now determine what is called a *power set.* A power set $\mathcal{P}X$ of a set X is the set of all subsets of X. X and $\emptyset$ are both elements of $\mathcal{P}X$. An axiom secures the existence of power sets.

4.3. Exercises.

(1) Explain why $\emptyset$ and X are both subsets of X.

4.4. Membership is a relation different in every way from identity. What about inclusion, by comparison? (i) Inclusion is reflexive, $X \subseteq X$. Inclusion is thus to be distinguished from a proper inclusion like $a \subset b$, for X is a subset of itself but not a proper subset of itself: $X \not\subset X$. (iii) Inclusion is transitive. If X is a subset of Y then every member of X is in Y. But all the elements of Y are in Z. So every element of X is an element of Z:

$$(X \subseteq Y \quad \& \quad Y \subseteq Z) \qquad \Rightarrow \qquad X \subseteq Z$$

But (ii) inclusion is not in general symmetric: $\{x, y\} \subseteq \{x, y, z\}$, but $\{x, y, z\} \not\subseteq \{x, y\}$. The special case is to be noted: if $X \subseteq Y$ and $X = Y$, then $Y \subseteq X$.

Clearly inclusion, membership, and identity are three distinct notions. And yet, in ordinary language, they are all captured "by the one overburdened verb 'to be.'"* So, we write

Leo is a Chicagoan;
A Chicagoan is an American;
Pope Leo XIV is the present Pope of the Catholic Church.

These uses of *to be* are now specified as the following:

Leo $\in$ the set of persons born in Chicago;
The set of natives to Chicago $\subseteq$ the set of Americans;
Leo XIV $=$ the present Pope.

5. Building Sets

5.1. In this section we investigate what is typically referred to as the "such that" locution, and tie it to a notation for building sets.

One can build a set by listing its members, but our logical understanding does not depend on that approach. One can also use a method of abstraction to build the set and then work with the abstract. The abstract is kind of like a diagrammatic identifier or picture-name.

A fuller locution involving the locution to which I just referred is "The set of all objects o such that $\ldots o \ldots$." Thus, the abstract assumes the virtual form of a kind of property or predicate. For this locution there is the notation:

$$\{o : \ldots o \ldots\}.$$

One way to think of this notation is as a way to formalize the price of admission to an event, as in: "To ride on this ride at Hershey Park, one must be *this* tall." The abstraction formalizes the admission qualification to belong to a particular set. It builds the set.

---

* [**115**], p. 182.

5.2. For example, the power set $\mathcal{P}X$ of X is the set of all the a such that a is a subset of X:

$$\begin{aligned}\mathcal{P}X &= \{\, a : a \text{ is a subset of } X \,\} \\ &= \{\, a : a \subseteq X \,\}.\end{aligned}$$

The admission qualification that any a must satisfy to be a member of $\mathcal{P}X$ is that 'a is a subset of X.'

6. Exercises

(1) Which of the following are true?
  (a) If $X \in Y$ and $Y \in Z$, then $X \in Z$.
  (b) If $X \in Y$ and $Y = Z$, then $X \in Z$.
  (c) If $X \in Y$ and $Y \subseteq Z$, then $X \in Z$.
  (d) If $X \subseteq Y$ and $Y \in Z$, then $X \in Z$.

(2) For the false statements in the exercise just above, provide a counterexample by specifying three sets by membership.

(3) Determine for each pair below if the first is an element in the second $\in$ or included in the second $\subseteq$.
  (a) $\{0\}$ $\quad$ $\{\{0\}\}$
  (b) $\{0\}$ $\quad$ $\{0, \{\{0\}\}\}$
  (c) $\{0\}$ $\quad$ $\{0, \{0\}\}$
  (d) $\{\{0\}\}$ $\quad$ $\{0, \{\{0\}\}\}$
  (e) $\{\{1\}\}$ $\quad$ $\{\{1\}, \{1, \{1\}\}\}$

(4) Find five sets V, W, X, Y, Z such that:
  $V = W$
  $X \subset W$
  $Y \in X$
  $Z \subset Y$

(5) Consider three sets $\emptyset, \{\emptyset\}, \{\{\emptyset\}\}$. Prove that no one of these is identical with the others.

CHAPTER 3

Set Building Operations

1. Equinumerosity

It was Georg Cantor who showed that sets outnumber objects. Consider sheep, for example, and sets of sheep.

Say a shepherd has a flock of them near a copse of trees. The shepherd has a good amount of rope, and taking one rope for each sheep, ties that sheep to one of the trees, allowing only one sheep per tree. At the end of the process, either there will be sheep that are not tied off, or there will be trees that have no sheep, or each sheep will be tied to one tree and no tree will not have a sheep tied to it.

The correlation process determines either that (i) there are fewer trees, (ii) there are fewer sheep, (iii) there are the same number of sheep and trees. In general, the idea here is of equinumerosity. To say that a set A is equinumerous with another set B is to say that there is bijective function f mapping A onto B. "Bijective" means the function is one-one

$$A \sim B \qquad \text{iff} \qquad f\colon A \xrightarrow[\text{onto}]{1-1} B.$$

Now replace the trees with sets of sheep.

The general question is, once one correlates each sheep with a set of sheep, and the process runs its course, will (i) (ii) or (iii) hold? Cantor proved that (ii) holds. For, once we have carried out the process of correlating sheep with a set of sheep, we can ask to check our correlation. How random was it? Is each sheep correlated with a set of sheep that contains it? For example, Baxter the sheep has been correlated with some set of sheep, but now can we properly claim that Baxter is a member of the set of sheep with which he has been correlated:

$$\text{Baxter} \mapsto \{\text{a sheep}\}.$$

Is Baxter an element of the set with which he has been correlated? Sure, maybe Baxter $\mapsto$ {Baxter}. But maybe Baxter $\mapsto$ {Bobby}. Let us stipulate a set abstraction: those sheep that have not been correlated with a set of sheep containing them as a member. That is an additional set of sheep beyond the correlation, so there are more sets of sheep than sheep.

2. Union, Intersection, and Separation

Set abstraction, the set building operation we studied at the end of chapter two, allows us to form additional helpful concepts. We have mentioned the empty set already. Since no object o satisfies the admission qualification $o \neq o$, the abstract $\{\, o : o \neq o \,\}$ equals $\emptyset$. The union $X \cup Y$ of X and Y is the set of all objects o that are either in X or in Y:

$$\begin{aligned} X \cup Y &= \{\, o : o \text{ is in at least one of these} \,\} \\ &= \{\, o : o \in X \quad \text{or} \quad o \in Y \,\}. \end{aligned}$$

The intersection $X \cap Y$ of X and Y is the set of all objects o that are in both X and Y:

$$\begin{aligned} X \cap Y &= \{\, o : o \text{ is in both} \,\} \\ &= \{\, o : o \in X \quad \& \quad o \in Y \,\}. \end{aligned}$$

Let us incorporate the standard symbol for cardinality and specify that $|A|$ is the cardinality of A. If so, is the following equality true or false?

$$\begin{aligned} 4 = |&\{\, o : o \text{ signed the Declaration of Independence} \,\} \\ &\cap \{\, o : o \text{ was born in Boston} \,\}| \end{aligned}$$

If $X \cap Y = \emptyset$, then X and Y are *disjoint.* Say the sets in $\{X_1, X_2, \ldots\}$ are *pairwise disjoint,* if the intersection of every pair of them is disjoint: i.e. $X_m \cap X_n = \emptyset$, for every X_m and X_n in the collection.

One might generalize these operations for any set which has sets for its elements. Two operations look inside these inner sets. The first operation, the union $\bigcup X$ of X, collects any members of the *inner sets* into a new set, with "inner sets" defined so that

$$\begin{aligned} \bigcup X &= \{\, o : o \text{ is an element of a member of } X \,\} \\ &= \{\, o : o \in x \text{ for some } x \in X \,\}. \end{aligned}$$

The second operation, the intersection $\bigcap X$ collects together into a new set the elements that are in all the inner sets:

$$\bigcap X = \{\, o : o \text{ is an element in every member of } X \,\}.$$

So,

$$X \cap Y = \bigcap \{X, Y\},$$

and

$$\bigcup \mathcal{P} X = X.$$

Consider the list of sets $X_1, X_2, \ldots$. The union of these sets is

$$\bigcup_n X_n = \bigcup\{X_n : n \in \mathbb{N}\}.$$

Axioms provide for the existence of unions and intersections.

Let me introduce one further useful idea by an example. If one has a pocket full of coins, one can sort them into pennies, nickels, dimes, and so on. Let us ponder the notion of separating out the pennies from the other coins in my pocket. There is the set B of coins in my pocket. I put all the coins on the table and separate out the pennies, all the x such that x looks copper, or has a copper color. A really cool idea we haven't mentioned is the *Aussonderung*, the operation that does that separating out with some set B. From the set it uses a formula like 'x looks copper' as a device of separation to form a subset A of B,

$$A = \{x \in B : x \text{ looks copper}\},$$

and more generally for some formula ϕ

$$A = \{x \in B : \phi x\}.$$

This provides us with a quick way to generalize the idea of all the coins in my pocket that aren't copper.

$$C = \{x \in B : x \notin A\}.$$

In any case when such a C is constructed out of A and B in this way, C is said to be in fact the *relative complement* of A in B:

$$C = B - A.$$

3. Relations

When we looked at unordered pairs, I said I would later introduce the notion of *ordered pair*. Well, that time has arrived. The idea is to define a set so that one can fix things such that

$$\langle x, y\rangle = \langle z, w\rangle \qquad \text{iff} \qquad x = z \;\&\; y = w.$$

This can be facilitated by the following definition:

$$\langle x, y\rangle = \big\{\{x\}, \{x, y\}\big\}.$$

It should be noted that the triple is reducible to the ordered pair:

$$\langle x, y, z\rangle = \big\langle\langle x, y\rangle, z\big\rangle,$$

and in general

$$\langle x_1, x_2, \ldots, x_n\rangle = \big\langle\langle x_1, x_2, \ldots, x_{n-1}\rangle, x_n\big\rangle.$$

Let us pause to dwell on this, for a Cartesian reduction can be made apparent. The idea of such a reduction is to make three dimensions into two, and indeed for any number of dimensions $n+1$ there is a Cartesian reduction of n dimensions.

This can be directly conceived in the following way, by taking the triple

$$\langle A, B, C\rangle$$

and eliminating the third dimension. Imagine for example without loss of generality that $A = \{a_1, a_2, a_3\}$, $B = \{b_1, b_2, b_3\}$, and $C = \{c_1, c_2\}$. Then we do not need the third dimension to comprehend $\langle A, B, C\rangle$, because one can capture the exact same set using indices. One must remember that $\langle A, B, C\rangle$, is a set of two ordered pairs

$$\{\langle\langle A, B\rangle, c_1\rangle, \langle\langle A, B\rangle, c_2\rangle\}.$$

Then we index the first ordered part at c_1 and the second ordered pair at c_2, to generate a pair of genuine ordered pairs:

$$\{\langle A, B\rangle_{c_1}, \langle A, B\rangle_{c_2}\}.$$

Like order, dimensionality somehow is a feature of sets, one might say.

There has long been a question about the reality of relations, which have been treated as second-class entities for as long as I can remember. But we do not have to worry about this question from the extensional viewpoint because the relation aRb is just the ordered pair $\langle a, b\rangle \in R$, where R is a set of ordered pairs. For example, consider pairs of lovers

$$L = \big\{\langle \text{Romeo}, \text{Juliet}\rangle, \langle \text{Tristan}, \text{Isolde}\rangle, \langle \text{Frederic}, \text{Catherine}\rangle, \ldots\big\}$$

and more generally,

$$L = \{\langle x, y\rangle : x \text{ loves } y\}.$$

On this definition, another other pair in L is $\langle \text{Juliet}, \text{Romeo}\rangle$, noting that

$$\langle \text{Romeo}, \text{Juliet}\rangle \neq \langle \text{Juliet}, \text{Romeo}\rangle.$$

Likewise, the pair $\langle \text{Lancelot}, \text{Guinevere}\rangle \in L$, but $\langle$Guinevere, Lancelot$\rangle \notin L$, for his love was unrequited.

A binary relation between X and Y in general is a set of ordered pairs that is a subset of the Cartesian product

$$X \times Y = \{\,\langle x, y\rangle : x \in X \ \& \ y \in Y\,\}.$$

One extremely important kind of relation are those relations that have only one value for each argument. These relations are known as functions. These are like black boxes that have a unique single output for every input. The functions are those relations R for which whenever $\langle x, y\rangle \in R$ and $\langle x, z\rangle \in R$, $y = z$.

4. Natural Numbers and Counter Sets

We want some nice set properties for the natural numbers. If we define the natural numbers as counter sets, we capture those properties. For example, the property that $1 \subseteq 2$, and for any counter set n,

$$n \subseteq n+1.$$

Of course, it is also the case that a nice property would be $1 \in 2$, and in general for counter sets

$$n \in n+1.$$

Thus, if we set

$$0 = \emptyset,$$

then we can replicate the successor operation on the natural numbers using the union operation on sets.

$$0+1 = 1 = \{\emptyset\} = \emptyset \cup \{\emptyset\};$$

moreover,

$$1+1 = 2 = \big\{\emptyset, \{\emptyset\}\big\} = \{\emptyset\} \cup \big\{\{\emptyset\}\big\},$$

and

$$3 = \big\{\emptyset, \{\emptyset\}, \{\emptyset, \{\emptyset\}\}\big\}.$$

For any natural number n, the *successor* of n, $n+1$ is identified with the $n+1^{th}$ set

$$n+1 = n \cup \{n\}.$$

The idea is that any natural number n is to be identified with a set X_n having n elements composed only of the sets upon which the previous natural numbers are based. These are the "counter sets":*

$$\emptyset, \{\emptyset\}, \{\emptyset, \{\emptyset\}\}, \ldots.$$

Another approach to the same concept is to first define what we call an *inductive set* as the set that contains the empty set and all the successor sets. One defines the natural numbers as any set that belongs to every inductive set. The Axiom of Infinity asserts that such a set exists. Then one can define ω as the set of all natural numbers:

$$\omega = \{\, x : x \text{ belongs to any inductive set.} \}$$

---

* [**86**], p. 247.

5. Foundations

5.1. Infinity. In one sense, the series

$$0, 1, 2, \ldots$$

represents a kind of infinity. Any element of this series, any one of these numbers, has a successor , and that goes on and on forever. This has sometimes been called the *potential infinite*. The ellipsis, the dots of leaving out, are in a formal series the dots that go on "and so on" suggesting the elements left out follow the same pattern, which in this case is the pattern of succession. We can write down for any finite number of objects $o_1, o_2, \ldots, o_n$ the set containing just those objects and no more $\{o_1, o_2, \ldots, o_n\}$. That set is not a potential infinity. It is strictly finite, having exactly n members for some $n \in \omega$.

The ellipsis can be utilized to represent the set of the natural numbers:

$$\begin{aligned} \mathbb{N} &= \{\, n : n \in \omega \,\} \\ &= \{0, 1, 2, \ldots\}. \end{aligned}$$

ω is the limit of the series of natural numbers. Again we use the ellipsis twice for the set of all integers:

$$\mathbb{Z} = \{\ldots - 2, -1, 0, 1, 2, \ldots\}.$$

Bertrand Russell spent a lot of time trying to prove infinity, then gave up when he realized that infinity is a supposition at best. The axiom of infinity says an infinite set exists. We can picture it as the set:

$$\{\, \emptyset, \emptyset \cup \{\emptyset\}, \{\emptyset\} \cup \{\{\emptyset\}\}, \ldots \,\}.$$

5.2. The Axiom of Choice. The Axiom of Choice, or AC, is about sorting. As Russell points out, if we are talking about a collection of shoes then one can in principle choose the left shoe in each pair. AC abstracts from that choice in two ways, first to the infinite case and then from shoes to socks. There are six or so equivalent ways to state the Axiom of Choice, and there are other versions of differing strengths. One extremely interesting type of choice is called the *Principle of Dependent Choices* (DC):

$$\begin{aligned} &\forall A \forall R (R \subseteq X \times X \wedge X \neq \emptyset \wedge \forall a \in A \exists b \in B(\langle a, b\rangle \in R \\ &\quad \rightarrow \exists f \in {}^{\omega}A \forall n \in \omega(\langle f(n), f(n+1)\rangle \in R)).^{*} \end{aligned}$$

Of the many nice features of this principle, formulated by Paul Bernays, one is that it is equivalent to well-foundedness. We mentioned above that to be well-founded is to be totally absent of infinite descending chains. The Principle of Dependent Choices is sufficient for the *Countable Axiom of Choice* (AC_ω), which states

---

* See [**56**], p. 132.

that *each countable set A has a choice function.* Of course, the proviso is that for all $a \in A$, $a \neq \emptyset$, lest there be nothing to choose.

One form of the Axiom of Choice is in terms of an infinite Cartesian product.* The infinite Cartesian product can be formed by using what is called an *index set.* An index set is used to refer to members of some set by a label. The idea is that I is an index set for X, if there is a function mapping I onto X

$$f\colon I \underset{\text{onto}}{\longrightarrow} X.$$

If the index set is ω then one can form a set of ω sequences (with domain ω) each of which for its ith term picks out a member of some set $G(i)$. A member of the Cartesian product is like a thread that runs through all the $G(0), G(1), \ldots$, picking out one member of each of those sets. This form of the Axiom of Choice says that such *threads* exist.

5.3. One Step Back from Disaster. The abstraction principle is a powerful tool. It allows us to consider collections of objects that do not qualify as sets because they lead to paradox. One might say that abstraction sets too low a bar, but another way to look at it is that we are able to ponder abstractions through admission qualifications that take us beyond the realm of sets. We rule out these collections because, if allowed, they lead to paradox. We will curtail them in a way that allows us to take one step back from the disaster of paradox, which I discuss below.

Consider syllables and the names of numbers. Obviously, we name numbers and every name has a certain quantity of syllables. The name for the number one has one syllable, the name for the number seven has two, and the name for the number eleven has three. Of course, different names are associated with the same number. 81 is identified as 'nine squared' and also as 'eighty-one' and the latter name has more syllables. Still we can ask about all the numbers that can be named in fewer than 23 syllables by any means. Clearly, not all the numbers $1, 2, \ldots$ will be captured by this cutoff. So, now consider all the numbers that cannot be named in fewer than 23 syllables. That is a set of numbers that will have a least member. So we have the notion of "the least member of the set of all numbers that cannot be named in fewer than 23 syllables." But then we have named a number that cannot be named. This antinomy, or something like it, was first observed by G.G. Berry, a librarian by profession.

Paradoxes of this sort can be blocked in set theory by stipulating that we admit only the sets that result from axioms. The typical axioms are the nine axioms of Zermelo-Fraenkel set theory plus the axiom of choice, ten total axioms.

---

* [**33**].

5.4. Russell's Paradox. Another antinomy worth mentioning is the one Russell wrote down in a letter to Gottlob Frege in 1902. He published the result in 1903. We have already considered the set of sheep and compared that with a set of sets of sheep. What about the set of sheep that do not shear themselves? "Wait, sheep never do that!" Okay, what about the Barber of Seville? This Barber, let us suppose, he shaves only those men who do not shave themselves. Does he shave himself? Indeed, he shaves himself if and only if he does not shave himself.

Russell's antinomy generalizes this paradoxical situation. We mentioned that sheep do not shear themselves and classes do not, in general, have themselves as elements. He considers the class of classes that do not have themselves as members.

$$\begin{aligned} w &= \{\, x : x \text{ is not in } x \,\} \\ &= \{\, x : x \notin x \,\}. \end{aligned}$$

Now consider whether w is an element of the set just defined, namely w:

$$w \in w \qquad \text{iff} \qquad w \notin w.$$

That is an antinomy, perhaps the paradigmatic example of such. Set theory handles it by the already mentioned distinction between sets and classes. Many collections are sets. For example, the collection containing only the empty set is a set. Any collection of sets qualifies as a class; however, some classes do not qualify as sets. These are called "proper classes." A proper class has too many members, in the sense of the correlation of sheep with trees, to be a set.

5.5. The Universe of Sets. The universe of sets V is a hierarchy built up in levels from the bottom by using the set building operations power and union. One begins with the empty set. One establishes that the bottom level of the hierarchy of sets is the empty set:

$$V_0 = \emptyset.$$

To build the sets we need, we use the power set operation $\mathcal{P}\emptyset = \{\emptyset\}$. That is how we build the next level, $V_1 = \mathcal{P}V_0$, and in general any successive level of the hierarchy of sets is defined as the power of the previous level:

$$V_{n+1} = \mathcal{P}V_n.$$

The idea is to capture any set at some level or another. Here we have

$$0 = \emptyset \in V_1,$$

and

$$1 = \{\emptyset\} \in V_2,$$

but what about the set

$$\{\emptyset, \{\emptyset\}, \{\{\emptyset\}\}, \ldots\}?$$

To capture that set the hierarchy needs a limit concept, the limit level V_ω, achieved by taking the infinite union of all those successive steps:

$$V_\omega = \bigcup_{n \in \omega} V_n.$$

The levels do not stop there, for we can consider successors of V_ω. For any successor ordinal α, set

$$V_{\alpha+1} = \mathcal{P}V_\alpha.$$

Where α is a limit ordinal, set

$$V_\alpha = \{\, V_\beta : \beta < \alpha \,\}.$$

On this suggestion, the ordinal numbers $0, 1, \ldots, \omega, \ldots, \alpha, \ldots$ constitute the backbone of the universe of sets. Up close, it might look a bit crooked, but viewed in large it straightens out.

Oops. I have not defined the rank of a set. For some set A, the rank of A is the lowest level of the hierarchy that has all of the elements of A. The rank of A is the least α such that $A \subseteq V_\alpha$, so that $A \in V_{\alpha+1}$.

That is the briefest of overviews on the hierarchy of sets. The proof structure of large cardinal axioms high above in the hierarchy is fascinating, well worth further discussion. I try to come back to this in an appendix. In what follows more immediately, our focus is on physical reality, as we keep our heads down and work away on a grand, old, and intractable, and by that I mean provably intractable, problem for scientific investigation, the continuum.

6. Exercises

(1) Assume $X \in \mathcal{P}Y$, Y includes X. Show that $\mathcal{P}X \subseteq \mathcal{P}Y$.
(2) Let $X = \{0, 1\}$ Using the definition of the ordered pair, deduce that $\langle 0, 1\rangle \subseteq \mathcal{P}X$
(3) Use *Aussonderung* to distinguish the "Wheat" pennies and the "Lincoln Memorial" pennies from the other coins in my pocket.
(4) Recall the counting set definitions from above. What is 3? Do 4 for extra credit.
(5) What is the rank of 2 in the hierarchy of sets? What about the rank of 3? Do you want to generalize?
(6) Show that the natural numbers are equinumerous with the positive integers.
(7) Show that $\mathbb{N} \sim \mathbb{Z}$.

CHAPTER 4

Continuum

This was the Noblest Roman of them all.
Shakespeare, *The Tragedie of Julius Caesar*, V.v*

1. The Dispute between Two Accounts

1.1. In the last few chapters, we did some work on sets. We looked at membership, inclusion, and identity before describing our set-building operation using the *such that* locution. Then we looked at other basic operations on sets. We also briefly introduced each of the ten axioms of Zermelo-Fraenkel set theory with the Axiom of Choice, or ZFC.

Now, we will put that model to use as we turn to examine the continuum. Before we can state exactly what the continuum is, we must wrestle with the fact that there are differing and competing accounts of the continuum.

There is a dispute between two accounts of the continuum. Let me tell a short story that illustrates some of the issues involved in said dispute.

1.2. When I go for a run along the Seine with a friend, time and distance pass by me in a continuous stream. But I mark the passage with each step I take. Each step is a discrete part of that continuous run. My experience of time and distance is unbroken. It flows by like the river running downstream to Rouen, but the sweeping second hand on my watch keeps time, marking the seconds in discrete units. We pass landmarks along the way, first Pont de la Tournelle, then Notre Dame, then Pont Neuf, Pont des Arts, and Pont Alexandre III. These are distinct stages, but we do not disappear and reappear between stages. We pass by them smoothly, without any gaps. Our destination, the Eiffel Tower, once it is in view, does not lurch closer to view.

Adjacent elements like the bridges differ perceptibly from each other. They mark the passage of the run. But any two consecutive instants on the run differ only imperceptibly. Still, we suddenly round a bend, and the Eiffel Tower comes suddenly into view. The run is a coherent whole, yet I have characterized it sequentially.

* Mark Antony refers to Marcus Brutus.

Although adjacent parts differ imperceptibly, the beginning and the end, not to mention each stage, are quite unalike each other.

Although informally set out, these are some of the main features of continua. These features coalesce into two prominent attributes. One is that the run is a whole stretch any part of which is a stretch. The other is that the run has distinct elements. These attributes do not seem to be in harmony. No two stages on the run are exactly alike. When we reach the Eiffel Tower, the run is over, but we saw it coming from a long way off. There are discrete points along the way and continuous stretches. The continuity of each stretch is like the continuity of the whole run, but the discrete aspect remains.

What is continuous, however, has no jumps. To be continuous is to hold together in an uninterrupted and unbroken progression. I propose that we take a closer look.

How are we to assess these conflicting appearances? Explication is a kind of analysis involved in building theories. Some say that it clarifies the meaning of words. In another tradition, however, it involves a theory-laden investigation of reality, not the meaning of words. The analogy that philosophical analysis is like an investigation in analytical chemistry sets out some basic features of this tradition, which begins with the method of logical construction. The analogy continues to work to this day, insofar as the analysis is into theoretical elements like definitions, axioms, and postulates. An explication in this tradition is a deduction from these theoretical elements.

As a result, the theory guides one in what to say about the logically constructed objects. Logical construction focuses on reference, but since the construction fits the theory, the logical construct does not match the original object. A philosophical analysis in this tradition eliminates that original problem in the theory with the theoretical construct. Bertrand Russell is the progenitor of this tradition of explication. His method of analysis by logical construction spans the twentieth-century philosophy of mathematics as its precursor. But my argument in what follows is more conceptual than historical. I set things out historically in order to follow a development that clarifies the main idea.

1.3. The explication of the continuum works through formal structures. At the most direct and simplest level, the continuum is the number line. An elemental explication of the continuity of the line begins with discrete points and defines continuous stretches of the line in terms of those points. There is an example in the next chapter of a progression that is based on discrete terms but comes together into a stretch. We will see other interesting examples of constructing the continuity out of the individuality.

Then again, a physical line has been said to have a unity that an arithmetical continuum does not have. Continuity has a different explication, one that is based on stretches and not on points, to capture this unity. On this explication, the continuum is a unified stretch, any part of which is a stretch. Since these accounts are not compatible with each other, and stretches are not elements, one begins to wonder whether or not the continuum has discrete elements.

Different theoretical constructions clarify and illuminate different senses of continuity. A consequence of explication is that different theories that fulfill the formal constraints of a particular analysis are all acceptable, even though they are mutually incompatible. This consequence raises a more general question of whether the subject of an investigation that leads to a variety of explications is a cluster or array of these. I argue that this solution, however open-minded, faces its own challenges.

The issue of what follows concerns the interchangeable, but mutually exclusive, analyses of the continuum. The structure of the chapter follows a classic rising and falling action that brings to the fore the sections on continuity and the main dispute. Section 2 is preparatory. It presents the main dispute and lays the groundwork for continuity by making explicit two fundamental notions, the number line (section 3) and separability (section 4).

2. Points or Stretches

2.1. We will see many different accounts of the continuum. We will try to lump them into kinds. Bertrand Russell quipped that some philosophers see the world as a "bowl of jelly," while others see the world as a "bucket of shot."* This is a puzzle about how the world is. One way to quickly distinguish the two is by divisibility. The jelly he has in mind would probably be a uniform gelatin, or it might be more like tapioca, because jelly is that way in England. Later, we will look at similar options, and others. Unlike a bucket of shot, a bowl of jelly does not have discrete atoms, by analogy. Clearly, however, no entity both has and does not have discrete elements.

More precisely, we will see below a quick summary approach to the continuum from the real numbers understood as sets, with reals defined not as sequences but as cuts. We would call that a Shot view. The dispute over the continuum involves separate reduction strategies. We can distinguish two basic reduction strategies, one reduction to indivisible *points* and the other reduction to divisible *stretches*.

---

* [**73**], p. 114.

Although the story of the twentieth-century philosophical analysis of the continuum begins here, the question is an ancient one. An analysis that resolves into discrete elements that compose everything, but are not themselves composed, promises to provide a complete understanding of the fundamental constitution of reality. In the nineteenth century, brilliant mathematicians devised mathematical analyses that filled the continuum, now newly understood as the number line, with densely packed points. This meant that cuts anywhere in the line are filled without a gap. These points can be understood as the real numbers. One began with ratios of whole numbers to construct the real numbers. Another partitioned the line into left and right.

These explanations, however brilliant, do not capture one prominent feature of natural processes, namely that we do not perceive jumps in time or space. The classic objection is that the arithmetical continuum, now understood as the real numbers, is discrete not continuous. For example, on the classic definition, f is *continuous* at a, if

$$(\forall\epsilon > 0)(\exists\delta > 0)(\forall x)(|x - a| < \delta \rightarrow |f(x) - f(a)| < \epsilon)$$

That is a minimalist account, one that I am inclined to assign to Weierstrass.

The elemental constituents of the arithmetical continuum would not seem to fit together in the way our perception of singular unified processes suggests. Even so, one who is inclined to the issue of our human access to the phenomena might see in the brilliant mathematical work a new aspect of the phenomena. The new aspect is a discrete approach favoring individual points, now identified as numbers. In the *Principles of Mathematics* Bertrand Russell reformulates this difference as an antinomy about elements of the continuum.*

2.2. When Russell introduces the method of logical construction, he transforms philosophical analysis. Logic is to be the foundation for a principled investigation of the continuum. The explication of the continuum poses an interesting test case for philosophical method, for even if an explication is artificial, it is not free of ontological consequences. This reveals a puzzling feature of the connection between the method of philosophical inquiry and the object of investigation.

From ancient times, it has been supposed that the method of inquiry should match the object of inquiry. But what does this mean? And how do we fix the criteria for "matching"? After all, since the late nineteenth century, mathematical investigations of

* [**111**].

the continuum usually presume the actual infinite. The most important exception is Brouwer's theory of the continuum. Hermann Weyl, who was critical of analysis, found this theory attractive.* Please allow me to delay consideration of this topic at present with the promise that I shall return to it in a later chapter, the penultimate chapter of the book. For the present, while we perceive no jumps in nature, we also approach problems step by step. These two issues affect how we understand the continuum.

2.3. Let us briefly survey Russell's approach in the *Principles of Mathematics.* He uses a *progression*, which is discrete but also *connected.* The key components of the progression are its infinity and the relation among the successive elements that connects them together.

Indeed, if the philosophical understanding of the continuum is to be clarified through logical constructions of the fundamental notions, two further steps are required, the clarifications of relations and infinity. These two fundamental notions are required to understand such things as the continuous unity of the line. While infinity has received much attention in the past century, it is often through relations that we gain understanding of the infinite.

Once the logical construction of the relation is seen as a formal chain of constructions, two questions leap to mind. The first concerns how the links of the chain are forged and connected one to another. The order of the links and the foundation of that order must be considered. The second concerns how long the chain is. These chains are progressions, based in a theory of series.

A progression is a fundamental series of type ω, ascending continuously inside a compact series containing it. A segment is a theoretical generalization of the real numbers. Real numbers are segments. There are four classes of segments, in each of which a progression or regression can be constructed that has a limit belonging to one of the four classes but not necessarily to the class that contains the progression or regression. Each of these classes is a perfect series going along (the number line) to the right or left. A perfect series (1) contains in every interval an infinite number of points, and (2) is one in which every point is a limit. For a series to be continuous in Cantor's sense, it must be what he calls *perfect* and *well-connected* or *cohesive* (*bien enchainè, zusammenhángend*).† To isolate it from other related notions, I call a perfect and well-connected series in Georg Cantor's sense a continuous form progression.

The progression provides the continuity necessary for the continuum in most respects. This is an appealing and apparently minimalistic idea. The leap to the infinite should not be seen to

* [**119**]. † [**111**], p. 288

represent a major shift, for there is a rule in the finite case and in the infinite case. What was to matter was the rule, not the number. The extent of the series is to be determined by the number of individuals in the interpreted domain. That number itself, the number of things that are, is a matter of science. With the proper rule in place, a progression can be finite or infinite and can be represented using a network of relations.

It must be acknowledged that a relational simulation of the progression requires more than first order logic. Set theory is also required for this analysis. The progression reflects Leibniz's idea of an analysis, namely an ordered series, which may be infinite in length and some of the elements of which may themselves be series of infinite length. Russell had argued that not all relations can be reduced to monadic predicates. He named this the main thesis of the doctrine of external relations. The very relations that cannot be so reduced, namely transitive asymmetric relations, he then reconstructs as ordered series, an outward expression of an inward connection.

Russell will later completely replace the logic of relations with the theory of types, an early version of which is already in Appendix B of the *Principles*. Type theory remains as an alternative to set theory for the foundations of mathematics, while category theory offers a third option. The theory of relations is a minimal foundation for logic. Russell adopts a predicative approach to relations in a series of steps from the *Principles* through an axiom of relations.

The guiding insight is that the theory should not presuppose anything more than a bare outline sketch traced by an ordered progression of forms, which in turn can be thought of as an ordinal chain and categorized according to length. Two reductive advantages are the using of limits in the definition and avoiding infinitesimal segments.

3. The Number Line

3.1. Different explanatory approaches to continuity will fit the bill so long as they meet the formal stipulation. One suggestion to encapsulate this situation is that the concept of the continuum is an array or cluster of these different senses. I will next argue that this encapsulation is suggestive and potentially misleading. The research program on the continuum still has not resulted in a perfect fit between method and object of inquiry in the following respect. Finite human beings do, of course, have rules and procedures as guides; finite beings do not actually go through all the steps.

In its simplest dimension, the continuum hypothesis is a statement concerning the number of points on the line, which it compares (by the relation of equinumerosity) to an infinite cardinal number. It equates the size of the continuum with the second infinite cardinal number. This hypothesis catches the attention of philosophers because the hypothesis cannot be decided in the axioms of set theory. Both it and its negation are consistent relative to the axioms of sets.

Kurt Gödel built L to show the axioms of set theory to be consistent relative to the continuum hypothesis, and to the axiom of choice.* A quarter of a century later, in 1963, Paul Cohen established that the continuum hypothesis is independent of ZFC, while the Axiom of Choice is independent of ZF. The continuum, seen in this light, is the *number line.* As such, it consists of the real numbers, which in turn are comprised of the rational and the irrational numbers. The rational numbers are ratios of natural numbers. They provide for the density of the continuum, as follows: *Between any two ratios there is a third.*

And so, the rational numbers are densely packed on the number line. The real numbers were defined by Cantor in terms of Cauchy sequences. Here we define them as Dedekind cuts. We define Dedekind cuts and then define the set of real numbers as the set that contains every Dedekind cut.

With any partition of the line into left and right, each real number is defined as the infinite set of ratios to the left, as the limit or collection of all of them. So a cut is not the empty set nor is it the set of all the rationals."Ratios to the left" can be more carefully expressed by saying that for any ratio in the collection, other ratios that are less than it are also in the collection. Because of the dense packing, another ratio can always be found stuffed in. More precisely, Dedekind cuts (i) are nonempty proper subsets of the rational numbers; (ii) are downward closed; (iii) have no greatest member.

One Dedekind cut is *less than* another if and only if it is a proper subset of the other. Again, for two real numbers r_1, r_2:

$$r_1 < r_2 \qquad \text{iff} \qquad r_1 \subset r_2$$

One ought to prove that this definition is adequate for a linear ordering on the reals. But for now a real number is the set of all the smaller real numbers:

$$r = \{\, s \in \mathbb{R} : s < r \,\}.$$

Continuity comes from the incorporation of limits. It is the irrationals that provide for a smoothly continuous line. The order of real numbers is not discrete because they are not ordered by

* [**56**], pp. 28–34 and 114–134.

succession. This permits measures of distance, stretches, on the number line.

3.2. "All well and good, that is some useful stuff; however, we began with the number line and the question about how many points there are on the number line. That whole line of investigation seems to have been dropped." Sorry, let me go back. It was Cantor who posed the question about how many points there are on the number line. That is according to what Kurt Gödel tells us in his reflective paper on Cantor's continuum hypothesis.

The answer to the question about the number of points on the number line is not straightforward. Let us begin by trying to simplify things a bit. I do not want to talk about all the numbers on the number line, but just a segment of it. We have seen just above how a real number can be defined as a subset of the continuum. Indeed, it is typical to think of the real numbers as:

$$\{\, x : x \subseteq \omega \,\}$$

Cantor stipulated that the cardinality of the set of the natural numbers is $\aleph_0$. So we have

$$\mathbb{R} = \{\, x : x \subseteq \omega \,\} = 2^{\aleph_0}$$

When Cantor asked about the number line, he was asking, "What is the cardinality of this set"? How many real numbers are there?

They call them real numbers for a reason. That is why this material is so important, because it is the juncture between mathematical reality and physical reality. We are looking at a reduction of continua to points, but that does not seem to be the intuitive approach. For Weyl and Brouwer, the intuitive approach is to reject the use of the infinite in the explanation of the continua, for the very important reason that we are talking about reality and we live finite human lives in a finite world. But they do not get all the nice results that people want when they talk about continua. So, we will need a sizable quantity of numbers to pack the continuum so that on the discrete analysis, there is a real anywhere you cut it.

To answer Cantor's question about the points on the number line, let us focus on just part of the number line instead of the whole thing, say just the open interval $(0, 1)$, by which I mean the real numbers between 0 and 1 and not including them:

$$(0, 1) = \{\, r \in \mathbb{R} : 0 < r < 1 \,\}$$

One might here ask, "How can you get away with that? I thought we wanted the whole number line, all the reals." Good question. I'll answer in three steps to establish that $(0, 1)$ is equinumerous with $\mathbb{R}$.

Step One. Begin with the number line. Imagine a circle floating above the number line between 0 and 1 such that one point on the circle touches 0 and one point on the circle touches one and the center of the circle is directly above the midpoint of the interval $(0, 1)$. Remove the top of the circle, so we are left with a semicircle. Now, as we did with the sheep and the trees, tie off each point on the interval with a point on the semicircle by drawing a straight vertical line from one up to the other. These sets of points are equinumerous.

Step Two. In this step, we will see that the set of points just described on the semicircle maps one to one and onto the set of points on the number line. Begin from the center point of the circle we imagined above. From it, draw rays outward through each point in the semi-circle. Each of these rays will hit the number line at a unique point. That is because we are not including rays from the center through the points at 0 and 1. Those would be parallel to the number line and do not intersect it on the Euclidean plane. Since every point on the number line is covered, these two sets of points are equinumerous.

Step Three. Equinumerosity is a symmetric and transitive relation. Since we have described a bijective function f from the interval to the semicircle, and another bijective function g from the semicircle to the entire number line, we conclude that there is a bijective function h composed of those two functions from the interval to the number line

$$h = g \circ f \colon (0, 1) \xrightarrow[\text{onto}]{1-1} \{r \in \mathbb{R} : -\infty < r < \infty\}$$

That answers the last question.

But there is still one component missing. This component concerns the cardinality of these sets. To fix this component in our understanding, we need one more resource, Cantor's diagonalization argument, which establishes that there are more numbers in the interval $(0, 1)$ than can be listed. We say that a set is *denumerable*, if all of its members can be listed, even if the list might be infinite in length. We also call such a list *countable*. The numbers on the number line between 0 and 1 are not countable. I will prove this, hopefully to your satisfaction, in Appendix **B**.

3.3. Exercises.

(1) Specify the functions f and g as above for Steps One and Two. Be sure they are composed correctly to produce h.
(2) Establish that equinumerosity is a reflexive, symmetric, and transitive relation.

3.4. There are many other senses of continuity besides that of Richard Dedekind. Russell writes, "The word continuity has borne among philosophers, especially since the time of Hegel, a meaning totally unlike that given to it by Cantor."* Russell wants to discuss "the mathematical meaning of continuity." And yet, as he immediately attests, even "in confining ourselves to the arithmetical continuum, we conflict in another way with common preconceptions." Of the arithmetical continuum, Poincaré says that this is a continuum made up of discrete individual elements, without the bonded unity that was to connect the whole thing together.

The "intimate bond" described by Henri Poincaré, and quoted at length in the chapter subsequent to this one, is a bond that unites individual elements. But even this way of looking at it presupposes a lack of continuity. The first idea is that of disparate elements united in a line, whereas what is wanted is to begin with a stretch of the line itself, not the point. A guiding idea about the continuum is that it has no "gaps," and points are discrete. What is continuous is supposed to connect continuously and be entirely unbroken.

4. Separability

4.1. Let us now review some standard considerations concerning continuity. Aristotle delves into the continuum and the nature of its unending divisibility in the *Physics* VI and elsewhere. Although there is inadequate room to argue so here in this essay, there is a deep resonance between the ancient discussions of philosophical method and of the continuum. In any case, Aristotle argues against the atomists that continuity is inextensible.† A continuous line does not have the "stigma" (point) of indivisibility. It cannot be composed of indivisible points. If a continuous line is composed of points, the points would have to be continuous or in contact, but indivisible points are neither.

In the *Physics* VIII he distinguishes two senses of "continuous distance" as he considers an argument by Zeno. In one sense, a continuous distance is divided into an actually infinite number of units; in another sense, the number of units is only potentially infinite. A traveler passes over a continuous distance in a "continuous motion" in the sense that the number of units is potentially infinite. If the number of units is actually infinite, the motion will be "intermittent." It requires jumps that cannot be traversed in a finite period of time. A finite magnitude does not need to contain an actually infinite number of units. Continuity on Aristotle's definition is a relationship between parts joined tightly together.

Appeals to Aristotle in current discussions of continuity are potentially misleading. They must be made with care. Continuity

* [**111**], pp. 346f. † 231a24–b18 and 263a4–b8.

is a property of geometrical figures and physical bodies for Aristotle and other ancient Greek thinkers. It is not that a triangle has a continuous area, but that it is a continuous area. No part of this area is divisible into a part that is not infinitely divisible. "Wait, is not continuity rather a property of space than of circles and triangles?" Yes, but the more familiar setting for the continuity of space and not for the continuity of geometrical objects in space is a modern approach due to Leibniz.

Gottfried Wilhelm Leibniz considers the nature of a continuous transition from a starting place to an ending place. If there is to be a continuous transition to a final terminus in a natural process, somehow that terminus must be factored into or included in the transition itself if there are to be no gaps. Otherwise, a jump would be required to make the transition. We perceive no such jumps in nature, he says, a version of the Principle of Continuity.

He introduces the differential as a variable approach to talk about differences. The infinitesimal is the idea of something that is not nothing but is smaller than any finite quantity. No matter how many times it is added to itself, it still remains smaller than any finite number. It is a quantity that cannot be assigned any finite number.

Although infinitesimals have been questioned by Berkeley and others, and although questions remain, the infinitesimal gained traction in the twentieth century due to nonstandard analysis. Here continuity is not derived from what is discrete; continuity in this theory is mathematically autonomous. This is a solid explication, yet an appeal to the infinitesimal is involved, revealing a remarkable aspect of the explication. A lesson to be learned from this case is that an effective explication can go forward even when conventions dictate otherwise.

While the continuum is held to be undivided, it does admit division without limit. A key to this phenomenon is that when one isolates any portion of the continuum, that portion should share these same features. Poincaré in our selected quotation above, mentions the unity of the continuum. Any part of it has the same continuity, that it is infinitely divisible all the way. If somehow the divisibility is to terminate, then the result is the reduction of a region to points, which have a different nature.

What remains when the continuum is pulled apart? This is the ancient question of its *separability*. If a model of the linear continuum is punctiform, and that number line is cut anywhere on analogy with a Dedekind cut, then there is to be a point at that juncture corresponding to a real number that marks the division. That real number belongs to a segment on one side of the division or the other.* For contrast, another theory uses classical analysis to partition the line into mutually exclusive disjoint intervals, these being separable. Once the partitioning is exhausted, when the continuum is cut, it comes apart into two parts. To argue that our real idea of the continuum is the separable continuum, and so to reject the division marked by a cut, is an interesting alternative approach, but hardly enough to force us to reject this continuum and the clear extensions that it provides.

4.2. Having glimpsed several approaches to the continuum, let us look more closely at one of these. Infinitesimals are perceived to be diametrically opposed to sets. There remains substantial debate about which approach to take. Take for example the nonstandard approach using infinitesimals. This can be understood if one begins with the classical real line and then constructs the universe of sets over it. That standard structure can then be extended to a nonstandard structure that is saturated in a way that makes it richer than the standard structure by realizing types to capture features of elements that are not defined in the base language. This provides more information about how they operate in the model. The richness of saturation is added to the collected universe in the following way. If there is a property that for any finite sub-collection is satisfied by an element of the collected universe, then an element of the universe satisfies this property for the entire infinite collection.†

This saturated universe satisfies a Transfer Principle, so that it satisfies the same first order properties as the standard universe. Transfer is reminiscent of the Principle of Continuity.

The final assessment seems clear. This remarkable structure is built using the theory of sets and founded on the real number line. The discrete has not been entirely abandoned and is somewhat in evidence in this view of continuity. Even so, an infinitesimal length is a stretch like the whole. The infinitesimal is a micro-continuum. The second salient point in the definition of the continuum is the denial of infinitesimal segments. Our puzzle is that one construction reduces the continuum to stretches while another rejects this idea.

5. Reduction

5.1. An explication of a term is a kind of investigation in which a logical construction replaces the term with a theoretical entity that bridges a gap in an explanation. In forging this new connection, there is no claim to be clarifying an idea held by users of this term. Instead, there is a certain perplexity about a term that is shown by the explication not to have been warranted.

* [**16**], sec. IV. † [**5**], pp. 260–261.

The constructed explication is satisfactory if it bridges a gap in a theory. As a consequence, artificiality in the construction is not necessarily a drawback, for the explication fills the lack of a theoretical connection.

The elements of construction fit the theory but do not necessarily preserve the original phenomenon being studied. In the opening of the preface to the *Principles* Russell states that "the explanation of the fundamental concepts" is in order to have the kind of acquaintance with them that we have "with redness or the taste of a pineapple." But what we are becoming acquainted with has already been modified for theoretical purposes. Reductions suffice because they meet the theoretical conditions.

Reductions are not incompatible with naturalism. The argument against reductionism is that our sensory inputs do not determine the answer to any formulation of the question about which reduction to use. They underdetermine any such answer. Since all theories reduce the original object to some theoretical entity, the secret is to see that it is the theoretical entity that the theory talks about. Yet, the reductions of the continuum we have seen are in conflict with each other. Any explication meeting the specified formal constraint is *right*, or right enough. This might lead to the view that a concept of what is being explicated is available after all. It is an aggregate of the various explications. This would be the idea that the continuum is a concept composed of all the various explications, even though these are in conflict with each other. This summation, however, is beset by problems, as we have seen.

Before I finish, one clarification is in order concerning the logical construction of the continuum. Let us take one last look.

5.2. The explication of the mathematical continuum might seem to reveal mathematics to be purely formal, empty of empirical content, and separate from natural science; however, for any theory with adequate formal structure, one can contrive some principled but arbitrary division into two realms that will satisfy a formal criterion for division. The cleavage of the mathematical from the scientific is appealing. The issue is how to specify the division. But this issue turns into an insurmountable problem. This paradox of analysis implies that appeals to meaning in the philosophy of mathematics are circular.

If, for example, a continuous form progression is thought to be a simply formal chain of constructions and nothing in itself, in the way the links in a chain require no third thing, then we would seem to have witnessed a picture of something of absolute form without content. The basic result of the paradox of analysis, however, is that any appeal to meaning is circular, so there is no self-standing argument between the mathematical and the factual. Upon philosophical analysis, mathematics has not been revealed to be stripped of content, after all.

But then no special realm of the mathematical continuum can be set off and secured from the rest of scientific theory once and for all. The reason is that, for any consistent formal system that includes enough theoretical structure for arithmetic, there are truths about the system that cannot be proved within the system until the system is extended; moreover, there are truths about the new extended system, if it is consistent, that cannot be proved in that system until it is extended. Nor can unintended models be ruled out.

6. Summation

There is a sort of general case for any argument for the reality of the mathematical continuum. The general strategy is that any attempt to sequester off a realm as vacuous, and name that the realm of mathematics, is circular. The general strategy holds for any formal systemic account of the continuum.

This line of thought, if clear, counts against a particular view of the mathematical continuum.

The outcome of our discussion in this chapter is that the explication of the continuum as the number line fits the canonical approach nicely. But so far the question of the cardinality of the continuum does not seem to have a solution.

And so, I do not think I can leave things as they are now. This section on reduction depends too much on new ideas that have not yet been carefully spelled out. But I am too tired to go on now. Let us return to that task in the next chapter.

CHAPTER 5

Logical Construction

1. Analysis

1.1. There has been a resurgence in interest in the topic of the continuum and its history.* This chapter discusses the logical construction of the continuum.

Russell's position is that an analysis by logical construction throws new light on the ancient debate over the discrete and the continuous. Let us assess that position. We began to explore the dueling reductions, one to the discrete and the other to the continuous, in the previous chapter. But we have to fill in some details on the nature of philosophy before we go further.

Philosophical analysis is like chemical analysis, or so we have been told. One source for Bertrand Russell's analysis of mathematics just before the beginning of the twentieth century was the investigation of the continuum. Before proceeding further, that historical claim needs to be established.

In "The Fundamental Ideas and Axioms of Mathematics," written in July and August of 1899, Russell was skeptical about the concept of infinity.† But soon after, and by the time he writes his 1899–1900 draft of the *Principles of Mathematics,* his skepticism about infinity has dissipated:

> He is much more sympathetic [to the infinite], and long sections of the work are concerned with Cantor. This change is doubtless due to his reading of [**10**] in July 1899.‡

Russell will later say that this is the "one major division in my philosophical work." It will lead Russell to explicate mathematical relations, discarding the old subject-predicate view of propositions, and to assume, albeit grudgingly, the axiom of infinity.

The story of the twentieth century philosophical analysis of the continuum begins with Russell, yet the question was an ancient one. An analysis that resolves into discrete elements that compose everything but are not themselves composed promises to give us a complete understanding of the fundamental constitution

---

* [**60**]; [**50**]; [**120**]; [**72**]; [**9**]; [**4**]. † [**108**], ch. 20; [**112**], ch. 1. ‡ Griffin and Lewis in [**108**], p. 262; cf. [**34**], pp. 878–920; [**103**], p. 11; [**108**], ch. 11.

of reality. Yet, natural processes are continuous both in time and space, so discrete constituents do not seem to combine into a singular unified process. In the nineteenth century, Dedekind and Cantor devised mathematical analyses that filled the continuum, now newly understood as the number line, with densely packed points such that it could be cut anywhere without a gap. Russell, who had always been inclined to emphasize the issue of our human access to the phenomena, saw in Cantor's analysis a new aspect of the phenomenon.

There were, then, for Russell two fundamental problems in understanding the continuum. The first concerns the relations between the elements and the larger structures into which they were to be compounded to provide for the requisite continuity. The second concerns the actually infinite number of elements that were required to make sure that the density was sufficient for any natural phenomena. These problems of the continuum concern the relational quality of the components and their infinite quantity.

The ancient philosophical debate over discrete and continuous approaches to the analysis of the continuum continues to this day, but developments in the twentieth century, largely in response to the mathematical advances of the nineteenth century, have sharpened and clarified certain details, so that we see the old question in a new aspect. Russell colorfully formulates the main question by the "one major division" we have already seen, between those who see the world as indivisible, the bowl of jelly contingent, and those that picture discrete atomic particles, the bowl of shot contingent.

Russell moved from idealism to realism in 1899 or so.* Thereafter, Russell soon came to see logic as the foundation for a principled investigation of the continuum. While much has changed since then, an explication as it is understood today maintains central features of what he calls a "logical construction." The term "logical construction" is first used explicitly in 1924, but Russell had been become engaged in this project early on, even before he wrote the *Principles.* What explication shares with logical construction is that what is produced, the constructed explication, is satisfactory if it bridges a gap in the theory opened by a defective conception, and thereby clarifies a connection. As a consequence, artificiality in the construction is not necessarily a drawback. The premier example of this phenomenon is the theory of descriptions.†

1.2. Analysis is transformed in the twentieth century when Russell introduces the method of logical construction in his work on the philosophy of the continuum and the infinite. Do the explications of mathematical relations and the infinite reveal that

* [**110**], p. 21; [**55**]. † [**104**].

mathematics is simply linguistic? Below we introduce several broad responses to Russell's realism. Russell intends the logical construction to be a minimalistic method to provide a theoretical simulation of mathematics. Later in the century a paradox will arise concerning the nature of philosophical methodology and its consequences for our understanding of mathematics. Although philosophers explicate continuity in different ways, these can be correlated.

1.3. The explication of the continuum poses an interesting test case for philosophical method, for even though an explication can be artificial, it is not free of ontological consequences. This reveals an interesting and puzzling feature of the connection between the method of philosophical inquiry and the object of investigation. Since ancient times, it has been supposed that the method of inquiry should match the object of inquiry. But since the late nineteenth century any mathematical investigation of the continuum presumes the actual infinite, as I have mentioned, with the one important exception being Brouwer's theory of the continuum.* Moreover, while there are no jumps in nature, finite human beings approach problems constructively, step by step. These two issues affect how we understand our interface with the continuum.

2. A New Approach to Analysis

2.1. Philosophical analysis was traditionally understood as comprised of analysis and synthesis, these being like two sides of the same coin. Analysis was a principled separation, a decoupling of the complex compounded matter into simple elements, and synthesis was the reversal of that method, a serial construction from the elements back up into their complex combinations.

At the beginning of the twentieth century, logical construction begins to emerge as a new approach. There is clear evidence of a fundamental shift from the traditional view of analysis and synthesis to analysis and logical construction in Bertrand Russell's thought:

> analysis and logical construction are not simply converse operations. Russell stresses that the objects discovered and distinguished in analysis are "real" as are their difference. Thus there is a constraint on the "choice" of definitions and primitive propositions with which to begin.... It is only their deductive relations, in particular their deducibility from the axioms of the theory, that are relevant to the success of

* [**34**], pp. 1169 and 1186–1197.

> a construction. Logical constructions do not capture all of the features of the pre-theoretic entities with which one begins.*

Let this quotation serve for us as a preliminary sketch of logical construction.

Russell's discussion of analysis develops over time, but his official doctrine of analysis is not our main topic. Our interest instead is in logical construction. For the record, mathematics is synthetic according to Russell at the time he begins to work on the project of logical construction. A reason for this is no doubt that axiomatic and deductive methods in geometry have been traditionally thought of as synthetic.† The argument here is that logical construction is not purely linguistic but is tied to reality. That tie, of course, has been seen to be a matter of degrees, and reality as discussed in twentieth century philosophy is something of a shifting target.

2.2. In the preternatural hours just before the twentieth century dawned, Russell had been preparing for a course he would teach on Leibniz.‡ With this exposure to the categories of Leibnizian analysis still rather fresh in mind, or so his story goes, he encountered David Hilbert and mathematical logic, and envisioned a new course of action using logical analysis. He sets this out in the three opening paragraphs of the preface to the *Principles.* In the first paragraph, he presents the logicist reduction. In the second paragraph, he states the second main object of his work, "the explanation of the fundamental concepts which mathematics accepts as indefinable. This is a purely philosophical task, and I cannot flatter myself that I have done more than indicate a vast field of inquiry." These fundamental ideas are class, deduction, logic, Arithmetic, finite, infinite, Geometry, and the Logic of Relations.

Over the next two decades, Russell will extend this project outward from mathematics to the external world and the analysis of matter.§ He calls it an analysis, but it is a completely constructive project. The division of relational structure into primitive indefinable elements is something like the process of investigation in analytical chemistry, insofar as both cases involve logical constructions from base elements. The loose analogy between mathematical and chemical analysis can then be extended to philosophical analysis. These cases involve logical construction. For the logical construction of the relation, accordingly, one begins with the relata and builds up the interconnections, as we are about to see.

---

* [**67**], sec. 2. † [**81**]. ‡ [**94**]. § [**107**] and [**92**].

2.3. Russell declares in his preface to the *Principles of Mathematics* that his two main objects are a logical analysis and a philosophical analysis. Both of these projects are constructive. The logical analysis is the technical work of the project that reduces mathematics to pure logic. Russell is already thinking about a theory of classes in connection with the foundations of mathematics, but he also associates mathematics with pure logic and perceives the extensional issues to be secondary to an intensional development through the theory of relations. Russell at this time does not see the theory of classes as a separate formal system. For him, logic is everywhere: "within each explicit formulation of logic all deductive reasoning, including all of classical analysis and much of Cantorian set theory, was to be formalized."* Russell says that his conception of the philosophical work being done is that it is not simply linguistic but, to the contrary, deals with the fundamental nature of reality. Indeed, we will see a couple of reasons to think that Russell was not so completely locked inside universalism as it might at first appear.

The other work is "a purely philosophical task," "the explanation of the fundamental concepts" so that we have the kind of acquaintance with them that we have "with redness or the taste of a pineapple," Russell says in the opening of his Preface.

3. Labyrinth

3.1. Russell came to think that if the philosophical understanding of the continuum were to be clarified through logical construction, two further steps were required, these being relations and infinity. Once the continuum is compared to a line, then relations and infinity are required to understand the nature of the line's continuous unity.

While infinity has received a lot of attention in the past century, it is often through relations that we gain understanding of the infinite. For example, to understand the actual infinite, one might explain mathematical induction and the concept of an inductive set. That concept involves the membership relation, truth functional operations, quantification, and succession. Succession is the operation that takes any natural number n and adds one to it. A proof by the principle of mathematical induction is successful if one can prove (i) that a property holds of zero, and (ii) that the property holds for $n + 1$ on the assumption that it holds for n.

Russell first states a version of that principle of induction as a primitive proposition in 1901. He soon, however, comes to see that

* Dreben and van Heijenoort in [**43**], p. 44.

as a "substantial error."* These are operational and relational ideas.

3.2. To advance his doctrine of internal relations, Bradley's idealistic monism required the assimilation of relations to monadic predicates. Russell had worked in this tradition through the end of the nineteenth century, when in 1900 he turned the argument around: the failure of the assimilation of certain relations to monadic predicates *is* the doctrine of external relations.†

Russell comes to reject the assimilation of relations to monadic predicates, because of the following transference problem: If Penny is taller than Henry, and Noah is taller than Penny, then Noah is taller than Henry; yet, this logical connection cannot be captured by a single monadic predicate. Relations like 'taller than,' which Russell would call asymmetric transitive relations, predicate different attributes of their terms. The problem is one of order, Russell writes.‡ The notation had, by then, been developed to capture the transitivity as: if xRy and moreover yRz, then xRz. This is the first chapter in the story of logical construction, the development of the logic of relations, as Russell will call it. Indeed, Russell will later define the doctrine of external relations as the position that not all relations xRy can be reduced to monadic predicates.

If the logic of relations is the first chapter, the second chapter concerns the class of all numbers and the paradox of the greatest ordinal. This leads Russell to his paradox, but also begins his attempt to construct the infinite from symbolic logic alone. Elementary arithmetic can be developed in set theory without infinity.§ But there was a problem, which Dedekind had anticipated toward the end of his career, namely that in the "arithmetic of infinite numbers," or any transfinite number theory, an axiom of infinity is "covertly invoked."¶

Cassius Jackson Keyser first recognized it. While Russell had originally disagreed with Keyser, in the second volume of the *Principia* Whitehead and he assume the axiom of infinity outright at $*$ 120.04. They follow Keyser in explicating the axiom of infinity in terms of an inductive class. With relations and infinity, one enters the labyrinth of the continuum.

Consider the Cartesian product of the continuum, $\mathbb{R} \times \mathbb{R}$.‖ A plane can be characterized as this Cartesian product by specifying

---

* 27 July 1901 letter to Couturat: [**112**], p. 313; cf. [**101**], sec. 4, Prop. 3.1. † Part IV, ch. 3 of the first draft of [**111**] in [**112**]; cf. Moore in sec. I of [**112**]; some monadic predicates are explicable as functors: [**83**]. ‡ [**112**], pp. 89–90; cf. [**38**]; [**101**]; [**99**], p. 171. § [**75**]. ¶ [**93**], p. 256; cf. [**17**]; cf. [**58**] and [**59**]. ‖ [**115**], p. 228.

a unit and the origin. Consider a straight line on this plane. It is a subset of $\mathbb{R} \times \mathbb{R}$. But then it is a relation.

3.3. Exercises.

(1) Explain how a straight continuous line can be thought of as a relation.
(2) Consider the real numbers and the relation $\leq$ for them. What is the converse of that relation?

4. The Progression

4.1. While the notion of logical construction is not exempt from ambiguity, some of the central debates surrounding it can be clarified by appeal to an ordered series or Dedekind chain to reconstruct the relation. If the logical construction of the relation leads to a formal chain of constructions, two questions leap to mind. The first concerns how the links of the chain are forged and connected one to another. The order of the links and the foundation of that order must be considered. The second question is how long.

Russell's *progression* is explicit in "The Logic of Relations with Some Applications to the Theory of Series."* A progression is a fundamental series of type ω, ascending continuously inside a compact series containing it. He says that the logic of relations "must serve as a foundation for mathematics, since it is always types of relations which are considered in symbolic reasoning." Toward the end of the paper, in the last few pages, Russell "advances the theory of segments, which constitutes a generalization of the theory of real numbers." Real numbers are segments on Russell's theory.

To review, there are four classes of segments, in each of which a progression or regression can be constructed that has a limit belonging to one of the four classes but not necessarily to the class that contains the progression or regression. Each of these classes is a perfect series "either to the right or to the left, that is to say, either for regressions or for progressions." A perfect series (i) contains in every interval ϵ an infinite number of points, and (ii) is one in which every point is a limit. A series is continuous on Cantor's view if it is *perfect* and *well-connected.*

4.2. At the end of the First World War, Russell advances some of these same ideas about ordered series from a more extensional perspective. Infinity and structures follow out of operations like $n+1$; similarities between relations themselves lead to structure. There is a tendency to overreach in interpreting Russell's remark:

---

* [**101**], §6.

> The method of 'postulating' what we want has many advantages; they are the same as the advantages of theft over honest toil. Let us leave them to others and proceed with our honest toil.*

Yes, Russell uses the word postulating here, but the issue at hand is not really the question of how seriously we are to take the axioms.

The question at hand is the theory of progressions once again. He will not merely postulate that there are external relations that cannot be internal, he is going to construct them. This quote is to be read in the context of his theory of progressions, because the work he is about to do is the toil of developing that theory. That theory is a way to reconstruct the relation, such that nothing but a formal theoretical connection is forged, while at the same time providing for asymmetric transitive relations, and thereby defying the doctrine of internal relations, on the one hand, while maintaining an early insight that a relation need not be a whole separate being unto itself.

The continuity necessary for the continuum can be generated constructively through the idea of a progression. The progression is fundamentally based on the Leibnizian idea of an analysis, namely an ordered series, which may be infinite in length and some of the elements of which may themselves be series of infinite length. Russell had argued against Bradley that not all relations can be reduced to monadic predicates, and named this the main thesis of the doctrine of external relations. The very relations that cannot be so reduced, namely transitive asymmetric relations, he then reconstructs as ordered series, which he intends to be an outward expression of an inward connection. Russell will later completely replace the logic of relations with the theory of types, an early version of which is already in Appendix B of the *Principles*.† Type theory remains an alternative to set theory for the foundations of mathematics, while category theory offers a third option.

4.3. The theory of relations is a minimal foundation for logic. Russell adopts a predicative approach to relations in a series of steps from the *Principles* through an axiom of relations. He writes:

> We assume then, that every function is equivalent, for all its values, to some predicative function of the same argument. ... we will call this

---

* [**97**], p. 71; [**8**]. † [**116**], 286–290.

> assumption the *axiom of classes*, or the *axiom of reducibility*.‡

It does not require anything more than the bare outline sketch traced by an ordered progression of forms, which in turn can be thought of as an ordinal chain and categorized according to *length*. Russell writes: "The salient points in the definition of the continuum are (1) the connection with the doctrine of limits, (2) the denial of infinitesimal segments."*

That continuity requires an infinite series is a mark of the phenomena under consideration. The object of investigation somehow escapes the methods of ordinary human experience. This is puzzling, because it is suggestive of the idea that mathematics is one step removed from reality, which then led several important twentieth century philosophers to take another closer look.

5. The Great 20th Century Ontological Debate

5.1. The generation of philosophers that immediately followed Russell wondered about the reality of mathematics. Russell's technical work commits him to this reality, but he intended to be sparing and minimalistic in his approach.† The discrepancy between Russell's intention and the outcome suggests that Russell is more concerned about the nature of philosophical reflection than about ontological commitments. Since they are formal and a step removed from experience, mathematics and logic seem to have a different character from natural science.

Phenomenological analysis, by way of contrast with the focus of the present paper, incorporates experience in a generalized way into the meaning of the term, so in this mode there are no "Russellian" thoughts about the external world that depend on objects.‡ Is there a way to separate out this realm from the realm of natural science?

5.2. The position that mathematics and logic are vacuous, entirely bereft of content, is a prominent feature of Carnap's dualism:

> By the time Quine met him, Carnap had adopted as his basic philosophical tenet what he took to be the fundamental insight of the *Tractatus* of Wittgenstein, namely, that *philosophy, logic, and mathematics have no content, make no truth claims.* Carnap saw *his* philosophical task to be the explication of this insight, to make it perfectly precise. The crucial function of analyticity for Carnap has always been to

‡ [**102**], p. 167. * [**111**], p. 353. † [**82**]; [**47**]; cf. [**49**]. ‡ [**54**]; [**53**]; [**76**], p. 200.

> articulate his conviction that philosophy and pure mathematics are not science.§

Carnap's development through his syntactic period can be read as responding to and increasingly rejecting the *Tractatus*, especially the Tractarian tripartite formulation.* In the *Tractatus*, logic and mathematics are "sinnlos." They are neither true nor false, and as such are unlike natural science, but are not "unsinnig," which is to say that they are not nonsense, and as such are unlike philosophy.† Only natural science is true in the *Tractatus*.

Unlike Wittgenstein, Carnap

> shifted 'tautology' from the category of *senseless* to the new philosophical category of *empty of material (factual) content (inhaltsleer), yet still having sense.*‡

He thereby separated the meaningful into two domains. Some meaningful statements hold because of their logical form and make no claim on reality. Examples of such statements include tautologies and mathematical equations. Those that do make a claim on reality, and state something truly or falsely, are the empirical statements of physics and natural science.§ Carnap then reformulates "tautological" as "analytic."¶

Quine, unlike Carnap, "is and always has been a faithful monist, not a dualist. Philosophy, logic, and mathematics are continuous with (natural) science."‖ For "Quine truth is truth and existence is existence," Dreben announces. Quine it should be noted gives a slingshot argument for monism that is something like Frege's. Church seems to have given a similar argument.

5.3. This completes a far too brief review of some of the ontological debates that arose out of the great logical analysis of mathematics done by Russell and Frege. For all the debate, the methodologies of Carnap and Quine have much in common.** Quine rejects Cartesian first philosophy, the idea that philosophy has an Archimedean point, as Descartes claimed in his Second Meditation, where he stated that without one "*firmum & immobile*" point to shift the world, philosophy for Quine cannot make a new beginning outside of or ahead of scientific inquiry. Before saying more on this topic, Quine's critique of meaning should be introduced.††

§ [**20**], pp. 85f.; cf. [**87**], pp. 129–132. * [**11**], pp. 8–9. † [**127**], 4.003, 4.11, 4.111, 4.461, 4.4611, 6.2, 6.21, 6.22. ‡ [**11**]. § [**15**], 76. ¶ [**14**]. For Carnap's dualism: [**12**]; [**13**]. For the analytic-synthetic distinction: [**82**], ch. II; [**78**]; [**55**], sec. 3). ‖ [**20**], p. 87. ** [**21**], pp. 4–5. †† [**82**], ch. II, sec. 6; [**88**]; cf. [**77**], pp. 128 and 151f.

6. A Critique of Meaning

Once tied to the nature of philosophical inquiry, more recently philosophical analysis is often considered to be synonymous with conceptual analysis, which in turn is usually associated with the probing of the meaning of words. The method of probing, however, has encountered substantial criticism over methodology and the status of the results. If the result of the probing is a new term with the same meaning as the original, then the claim to have explained something in an informative way, or to have clarified something that was conceptually unclear, has not been met.

This, according to Quine, is the "paradox of analysis":

> Philosophical analysis, explication, has not always been seen in this way. Only the reading of a synonymy claim into analysis could engender the so-called paradox of analysis, which runs thus: how can a correct analysis be informative, since to understand it we must already know the meanings of its terms, and hence already know that the terms which it equates are synonymous?*

Explication does not separate the investigation from reality. "Explication is elimination," Quine explains. But what does he mean by that?

When one introduces a new approach that avoids a vague or troublesome expression, one is not merely substituting one word for another, which would at most constitute a transformation of language. An explication of a term is a kind of investigation in which a logical construction replaces the term with a theoretical entity that bridges a gap in an explanation. In forging this new connection, there is no claim to be clarifying an idea held by users of this term. Instead, there is a certain perplexity about a term that is shown by the explication not to have been warranted. It is the problem that is eliminated, by banishing "needless usages."

The ordered pair, for example, is the device that fixes talk about relations by allowing the precise mathematical combination for comparison of two disparate elements. The explication of the relation by the ordered pair not only reveals that the relation can be simplified further, since by definition a triple or sequence of any number of elements can be reduced to an ordered pair, but also that the ordered pair is an (unordered) set containing the pair and a singleton containing the first of the pair, which we have studied in **3.3** above:

$$\langle x, y\rangle = \{\{x\}, \{x, y\}\}.$$

---

* [**88**], pp. 259f.

Russell says that order is fundamental to relations. But then might we not say that the following equality proves him wrong?

$$\{\{x\},\{x,y\}\} = \{\{x,y\},\{x\}\}$$

Here again is evidence of the fecundity of the set.

I am using the Kuratowski version of the ordered pair. The Kuratowski version works because of the unspoken convention that x is the first term and y the second. Another definition is to use two old-fashioned symbols, one for the empty set Λ and one for its dual, V, the *set* of all sets, or we should say the *class* of all sets, as follows:

$$\langle x, y\rangle = \{\{x,\Lambda\},\{y,\mathrm{V}\}\}.$$

Norbert Wiener first simplified the logic of relations through the ordered pair.*

Russell, however, might be said to have got the ball rolling on this one. He is the first I have identified to consider the relation in extension as a class of couples.†

Quine, by the way, argues against reductionism as one of the two dogmas of empiricism, by which he means to advance his holism.‡ The word 'reduction' as used in the Cartesian reduction above is not to be understood as implying that the statement should be considered in isolation from the theory. The reduction from a sequence of length $n+1$, $\langle x_1, x_2, \ldots, x_n, x_{n+1}\rangle$, to a sequence of length n is accomplished in a way we have already seen. A sequence of any length can be reduced to an ordered pair. Quine urges us to beware the dogma of reductionism, the favoring of one explication over another, but obviously reductions are harmless in and of themselves. One would want to remind him that any explication involves a new reduction of some sort.

Quine says that the construction of the ordered pair

> is paradigmatic of what we are most typically up to when in a philosophical spirit we offer an "analysis" or "explication" of some hitherto inadequately formulated "idea" or expression. We do not claim synonymy. We do not claim to make clear and explicit what the users of the unclear expression had unconsciously in mind all along. We do not expose meanings, as the words 'analysis' and 'explication' would suggest; we supply lacks.§

As a result, one comes to think about order in a different way. An order between a pair of two things requires that attention be called, perhaps by an isolated repetition, to the first of the two.

---

* [**124**]. † [**111**], p. 99. ‡ [**82**], ch. II. § [**88**], sec. 53.

The reader is invited to compare Quine's account of explication with the preliminary sketch of logical construction above. But perhaps the preliminary sketch is retroactively influenced by Quinean explication.

To answer such a retort, the reader should consult Russell's own words on the philosophy of the continuum, as follows:

> The notion to which Cantor gives the name of *continuum* may, of course, be called by any other name in or out of the dictionary, and it is open to every one to assert that he himself means something quite different by the continuum. But these verbal questions are purely frivolous. Cantor's merit lies, not in meaning what other people mean, but in telling us what he means himself—an almost unique merit where continuity is concerned. He has defined, accurately and generally, a purely ordinal notion, free ... from contradictions, and sufficient for all Analysis, Geometry, and Dynamics. This notion was presupposed in existing mathematics, though it was not known exactly what it was that was presupposed.*

The parallel between Russell's presentation of Cantor's merit and lucidity and Quine's discussion of explication suggests to me that explication stems from logical construction. One might insist that what is accomplished is no more than a translation, but that collapses a distinction between description and construction. An explication gives an explanation and has consequences that pure description is not supposed to have.

* [**111**], p. 353.

CHAPTER 6

Natural and Scientific

1. The Purported Distinction between Analytic and Synthetic

Let us return to the appealing idea of a division of a theoretical realm into analytic and synthetic. When Quine presents the paradox of analysis, his point is that one can always look at any serious theory with adequate formal structure and insist on some principled but somewhat arbitrary division into two regions that will satisfy Carnap's criteria for division, which implies that appeals to meaning in the philosophy of mathematics are circular.

The explication of mathematics would seem to reveal mathematics to be purely formal, empty of empirical content, and separate from natural science. If a construction chain like a continuous form series is thought to be simply formal and nothing in itself, in the way the links in a chain require no third thing, then we would seem to have witnessed a picture of something of absolute form without content.

The basic result of the paradox of analysis, however, is that any appeal to meaning is circular and so there is no self-standing argument for the distinction of the mathematical from the factual. Upon philosophical analysis, mathematics has not been revealed to be stripped of content, after all. But, then, no special realm of mathematics can be set off and secured from the rest of scientific theory once and for all.

There you have an argument against the position that all of mathematics is analytic. The continuum, for example, is not a creature of syntax. Please allow me to point out that this line of debate is separate from another question about Carnap's philosophy, namely to what extent the inherent incompleteness of any formal system damages other features of his developing position.*

There is a sort of general case tactic for any such argument for the reality of mathematics. The general tactic is to reject any attempt to sequester off a realm as vacuous, and name that the realm of mathematics. The rejection is because the attempt itself is circular. Hilary Putnam added a further dimension to

* [**77**]; cf.[**3**]; [**46**].

the problem. Unintended models cannot be ruled out. If there is any model of set theory at all, then there is also an unintended model.* The general tactic holds for any formal system, that is any system adequate to elementary arithmetic.

This clear line of thought counts against Russell's later view of mathematics. Russell was influenced by Wittgenstein's views of mathematics—as he understood them; "once having accepted them, he never abandoned them in his later years."†

Russell eventually argues that the valid propositions of truth functional logic, or tautologies, "can be known without appeal to external observation."‡ In one stroke, he adds to this core all "mathematical and logical propositions," since the form of any mathematical proposition can be explicated so as to exhibit basic relational structures. Once generalized, "we are enabled to speak about relations." For comparison, to speak about a relation in set theory is to speak about a set of ordered pairs. But he goes further, jumping to the conclusion that the "propositions of logic and mathematics are purely linguistic, and they are concerned with syntax." But that conclusion depends upon the "further property needed to make a proposition one of mathematics or logic." While this property, he acknowledges, is "difficult to define," it "is the property traditionally expressed by saying that the propositions concerned are 'analytic.'" But to define that property is not only difficult, it is ineluctably circular. Russell, it seems, gave up on realism too quickly.

Quine takes a different approach to the analytic, which he externalizes, domesticates, and incorporates into his system as those propositions that will command universal assent.§ But his rejection of first philosophy forbids that such a realm can be sequestered once and for all. "No statement is immune to revision."¶ Quine's holism dictates that experience relates not to individual sentences but to sets of true theoretical sentences that are subsets of the theory and in principle to whole theory. There are many relevant recent theories and much current debate.

In conclusion, there are consequences to the explication of a mathematical concept that are not purely linguistic and go beyond the manipulation of expressions by rules. Finite human beings do, of course, have rules and procedures as guides; finite beings, however, do not actually go through infinite steps in the process of mathematical proof. All the progress in the philosophy of mathematics in the last twelve decades, then, has not resulted in a perfect fit between method and object of inquiry, particularly when

---

* [**79**]; cf. [**80**]; [**57**], sec. 3. † Lackey in [**96**], p. 16. ‡ [**98**], pp. 303–306. § [**85**], pp. 77–80; [**88**], pp. 65–69; [**87**], pp. 127–130. ¶ [**82**], p. 43; cf. [**21**], ch. 8; [**50**], sec. 3.1.

it comes to relations and infinity, which lead us, as Leibniz says, into the "labyrinth of the continuum."

2. Continuity

2.1. There are many senses of continuity. Russell writes, in a quote we have already seen, "The word *continuity* has borne among philosophers, especially since the time of Hegel, a meaning totally unlike that given to it by Cantor." But Russell wants to discuss "the mathematical meaning of continuity"; however, as he immediately attests, even

> in confining ourselves to the arithmetical continuum, we conflict in another way with common preconceptions. Of the arithmetical continuum, M. Poincaré (*Revue de Métaphysique et de Morale*, 1: 26) justly remarks: "The continuum thus conceived is nothing but a collection of individuals arranged in a certain order, infinite in number, it is true, but external to each other. This is not the ordinary conception, in which there is supposed to be, between the elements of the continuum, a sort of intimate bond which makes a whole of them, in which the point is not prior to the line, but the line to the point. Of the famous formula, the continuum is unity in multiplicity, the multiplicity alone subsists, the unity has disappeared.*

The "intimate bond" described by Poincaré is a bond uniting individual elements.

But even this way of looking at things, he seems to be saying, presupposes a lack of continuity, because the first idea is the idea of disparate elements united in a line, whereas what is wanted is to begin with a region of the line itself, not the point, for a guiding idea about the continuum is that it has no "gaps," and points are discrete. What is continuous is connected continuously and is entirely unbroken. The use of "arithmetical" is to suggest the discrete as opposed to the continuous in the sense of geometry.

Russell states the antinomy of the continuum as follows: "The denial of infinitesimal elements resolves an antinomy which had long been an open scandal, I mean the antinomy that the continuum both does and does not consist of elements."† Russell's antinomy raises again an ancient topic of debate.

* [**111**], pp. 346f. † [**111**], p. 353.

2.2. Aristotle delved into the continuum and the nature of its unending divisibility. It can be argued, although that is not done here in this book, that there is a deep resonance between the ancient discussions of philosophical method and of the continuum.

Aristotle says that the essence of a finite magnitude does not include that it contains an actually infinite number of points or regions.* According to Hellman and Shapiro, Aristotle's continuum satisfies the Archimedean principle, and continuity is a necessary condition of infinite divisibility.† Aristotle does not define the continuum but instead works relationally from parts "joining up" or being "glued" together.

Strictly speaking, the appeal to Aristotle in current discussions of continuity is potentially misleading. Continuity is a property of bodies or geometrical figures for Aristotle and the ancient Greek thinkers.‡ It is not that a triangle has or possesses a continuous area, but that it itself *is* a continuous area.

An expert on Leibniz reports that

> Aristotle construes continuity as a relation among successive parts of a magnitude, not as a structure arising from relations among all the possible parts of a whole, as does Leibniz. ... [The] notion of continuity as a property of *space*, as opposed to the figures or bodies that might be in it, was a seventeenth-century innovation, issuing mainly out of Leibniz's studies of *Analysis Situs*.§

Leibniz thought deeply about the nature of a continuous transition from a starting place to an ending place.

If there is to be a continuous transition to a final terminus in a natural process, somehow that terminus must be factored into or included in the transition itself, if there are to be no gaps. Otherwise, a leap is required to make the transition, but nature makes no such leap: "*natura non facit saltus*," Leibniz says many times, this being the Principle of Continuity.¶

Leibniz introduced the differential as a variable approach to talk about differences. The infinitesimal is the idea of something that is not nothing but is smaller than any finite number: no matter how many times it is added to itself, it still remains smaller than any finite number. This is a quantity that cannot be assigned any finite number.

While infinitesimals have been questioned by Berkeley and others, the infinitesimal gained traction in the twentieth century due to nonstandard analysis. Here continuity is not derived from

---

* *Physics* VIII, 263b3ff. † [**50**], pp. 1f. ‡ [**18**]. § [**2**]. ¶ [**65**].

what is discrete; continuity in this theory is mathematically autonomous. This is a solid explication, yet an appeal to the infinitesimal is involved, revealing a remarkable aspect of the explication. An effective explication can go forward even when conventions dictate otherwise.

Put more directly, there are different dimensions of the continuum. It can be seen as a plane, or a line, for example, as Cantor and Dedekind discussed in a series of letters before the beginning of the twentieth century.* Since then, many different conceptions of continuity have been introduced. A recent survey of the main options by Reeder divides the "plurality of philosophical conceptions of continuity" into two forms of explication, those for which the parts are prior to the whole and those for which the whole are prior to the parts.†

While the continuum is held by some to be undivided, it does admit division without limit. If one isolates any portion of the continuum, that portion should share these same features. Poincaré, in our selected quotation above, mentions the unity of the continuum. Any part of it has the same continuity, that it is infinitely divisible all the way. The unity contains within it a plurality. If somehow the divisibility is to terminate, then the result is the reduction of the line to points, which have a different nature because they possess one fewer dimension.

And yet, no matter how persuasive that deeply metaphysical view might be, it does not resonate with me. As I have already taken pains to point out, more than once, a dimensional reduction like a Cartesian reduction of the cube to the plane is said to involve a change in nature. That is a metaphysical way to look at the thing. I do not look at it like that.

I look at it the other way around. Why? Because I am not beginning with an assumption about how our reality should be, one, two, or many. Instead I am beginning with an assumption about philosophical and natural analysis. So, I begin to investigate the continuum by facing the problem of order. I then turn to the representation of order as the series, and then the ordered pair. do not see a change in nature. "Tell me, once more, why that would be."

Because I am beginning from the formal exposition perspective. I am working with physical objects like sets of real numbers. The cube is a physical object, and so is the plane. One can work with sets of reals, and as we have seen, one can order sets of reals into n-tuples for the representation of the objects of geometry. But the series of reals we have seen are sets of sets of reals. That was an exercise in an earlier chapter.

* [**34**], ch. 19. † [**90**].

The formal appearance is of real numbers all the way down to sets. We defined them as subsets of counter sets. That leads me to observe that I have a perspective on reality rather different than that of M. Poincaré. I hope that is not completely true, because I did a lot of work on the continuous form progression to make a bridge to his view.

Having broken in with this interlude, let me now return to the main theme.

What remains when the continuum is pulled apart? This is the ancient question of its "separability," for which see chapter **4** above. If a model of the continuum is punctiform, and you cut that number line anywhere, on analogy with a Dedekind cut, then there has to be a point at that juncture corresponding to a real number that marks the division, and that real number belongs to a segment on one side of the division or the other. Here the continuum is like points packed densely together. But another model of the continuum is that when you cut it, it comes apart into two parts, as in a recent construction of the classical continuum without points.* Using classical analysis, the line is exhaustively partitioned into mutual exclusive disjoint intervals, these being separable.

2.3. We have glimpsed three approaches to the continuum, one from set theory and two from category theory. For the purposes of juxtaposition, let us focus on just two. There remains substantial debate about which approach to take. Infinitesimals are perceived to be diametrically opposed to sets. Such debates continue, but the categories can be seen in the light of sets. Take for example the nonstandard approach using infinitesimals. This can be understood if one begins with the classical real line and then constructs the set theoretic universe over it. That standard structure can then be extended to a nonstandard structure that is saturated in a way that makes it richer than the standard structure. This is done by realizing types to capture features of elements that are not defined in the base language. These new features provide more information about operations in the model.

The richness of saturation is added to the collected universe in the following way. If there is a property that for any finite subcollection is satisfied by an element of the collected universe, then an element of the universe satisfies this property for the entire infinite collection.† This saturated universe satisfies a Transfer Principle, so that it satisfies the same first order properties as the standard universe. Transfer is reminiscent of Leibnizian continuity.

---

* [**50**]. † [**5**], pp. 260f.

Continuity can be seen from the perspective of regions instead of points. These very different aspects of the continuum can be correlated with each other. The unity of the continuum suggests that its constitution is more like jelly than pellets of shot. But, even if one approaches it categorically, the continuum still has a discrete resolution. At last, any account is regimented into the canonical formulation. On the shot analogy, the question becomes how many pellets of shot are required to pack the continuum. That is an analogy for the hypothesis posed by Cantor.*

Now thinking of the continuum in perhaps its simplest respect, the continuum hypothesis is a statement concerning the number of points on the line, which it compares by the relation of equinumerosity to an infinite cardinal number. It equates the *size* of the continuum with the second infinite cardinal number $\aleph_1$. This hypothesis catches the attention of philosophers because the hypothesis cannot be decided in Zermelo-Fraenkel set theory. Both it and its negation are consistent relative to the axioms of that theory.†

3. The Constructible Hierarchy

With the progression, we begin to have a kind of bare minimal structure for mathematics. The progression extends to an infinite and continuous construction chain. The idea is a kind of ramification, like the process Russell envisioned.‡ The idea is to incorporate the properties in stages:

> the concept of the ramified hierarchy has been an important, if usually unremarked, influence in mathematical logic. For instance, in the 1930s Gödel discovered that, if the ramification of properties over the natural numbers is extended into the transfinite, the process breaks down at level ω_1, the first uncountable ordinal. (This may be considered a provable form of the Axiom of Reducibility.) This led him to introduce the hierarchy of constructible sets which he employed to prove the relative consistency of the continuum hypothesis.§

The constructible hierarchy L is a construction based only on sets that are first order definable.¶ In one sense, this is an extensional representation of the transfinite extent of order. It is a sparse conception, but it is decisive.

The position that mathematics is analytic could cover something like the claim that the universe of sets includes all the first order sets that are definable relative to the preceding sets in the

---

* [**48**]. † [**60**]. ‡ [**102**]. § [**6**], sec. 1. ¶ [**45**].

hierarchy and no more ($V = L$). The idea was of a very fine structure of threadlike chains of construction, all held in place by a transfinite ordinal backbone, and this very slight thing being all analytic. This is an appealing idea, and seemingly minimalistic. The leap to the infinite was not intended to represent a major shift, for we have a rule in the finite case and we have a rule in the infinite case, nor was the reality supposed to matter, on this view. What was to matter was the rule. The extent of the series is to be determined by the number of individuals in the interpreted domain. But that number itself, the number of things that are, was thought to be a matter of science. With the proper rule in place, the form series can be finite or infinite, although for the continuum, as we have seen, the actual infinite is a prerequisite for most theories.

Let us now take a last look at the big picture supposition, which Russell formulated as the position that mathematical propositions are purely linguistic and concerned only with syntax. Gödel thought of L as a transfinite extension of Russell's theory of types. If so, if what it means for all of mathematics to be analytic is to be something like the hypothesis that the mathematical realm is constructible through precise axiomatization and definition carried into the transfinite, then Russell's formulation extensionally involves something like a model of set theory in which the continuum hypothesis is true, which decides the continuum hypothesis in disregard of the consensus on its absolute undecidability. Whether or not one is comfortable with this outcome, the introduction of the axiom of constructibility would beg the question. Exercise (3) below asks for an explanation why. This result fits a conclusion we want to draw about explication.

There remains the question of how to resolve the antinomy of the continuum. Russell's solution to the antinomy is of the form in which both positions hold true, but in different senses:

> the continuum both does and does not consist of discrete elements. We see now that both may be said, though in different senses. Every continuum is a series consisting of terms, and the terms, if not indivisible, at any rate are not divisible into new terms of the continuum. In this sense there are elements. But if we take consecutive terms together with their asymmetrical relation as constituting what may be called ... an *ordinal* element, then, in this sense, our continuum has no elements. If we take a stretch to be essentially serial, so that it must consist

> of at least two terms, then there are no elementary stretches; and if our continuum be one in which there is distance, then likewise there are no elementary distances. But in neither of these cases is there the slightest logical ground for elements.*

On the one hand the continuous chain series is composed of individual ordinal terms, these being links in the chain that are not divisible into new links; on the other hand it is composed of series of consecutive terms that are themselves possibly infinite, linked together by the connecting relation: each of these stretches of the chain constitutes ordinal elements that have a non-elemental unity.

3.1. Exercises.

(1) Look up and write down the definition of L. The definition is in stages by transfinite recursion over the ordinal numbers.
(2) What is the universe of L?
(3) Explain why the *analytic* cannot be conclusively identified with, or clarified by, the axiom of constructibility.

4. The Antinomy

There is a conflict of senses, not only about the continuum in general, but about the arithmetical continuum. The antinomy, that the continuum has elements and does not have elements, Russell called an "open scandal." When you divide a continuum, are there two things, two stretches, or is there one thing, a continuum with a particular dividing point?

How does the philosophical community assess this situation? A recent survey of the main options splits the 'plurality of philosophical conceptions of continuity' into two forms of explication, those for which the parts are prior to the whole and those for which the whole is prior to the parts. The first approach features elements, while the second has no elements. On the sand analogy, the question becomes how many grains of sand are required to pack the continuum. That is an analogy for the continuum problem. Yet, the unity of the continuum suggests that its constitution is more like that of a gelatin or tapioca than like sand. Our concern in this section is that these are mutually incompatible theories. Different structures are required to explain the continuum as one mathematical concept or another.

One solution to the antinomy is of the form in which both positions hold true, but in different senses that relate to entirely

---

* [**111**], p. 353.

different theoretical aspects. This solution depends on the continuous form progression. We saw it in the quote from Russell at the end of the previous section. The ordinal terms are elements; stretches, however, are the basic units of the theory. Each stretch is a part of the continuum that has a non-elemental unity.

This resolves several key aspects of the antinomy. The resolution into continuous regions matches the concept of the continuum with powerful intuitions about the decomposition of the continuum, namely that parts of the continuum are themselves non-elemental and have continuity. It accommodates non-elemental regions by taking consecutive terms bonded in asymmetrical connection. A resolution into points matches the concept of the continuum with powerful intuitions about the ways of human knowing, that it is discrete and happens step by step. This option is sacrificed if one resolves Russell's scandal by a reduction to stretches. This dialectic is reminiscent of the situation we encountered above with the infinitesimals based in a formal theory of sets built upon the assumption of the real number line. There are elements involved in the foundational base assumption, but there are no elements involved in the ultimate reduction.

The ordinal elements are among the formal ingredients of the construction, but the formal structure of the theory takes the region as basic. This theory is not a reduction to elements, in other words, even though it has an ordinal basis. The main objection is that its minimalism forces a restricted picture of mathematical reality. I do not want to move on, however, without adding one thought about the lack of an intimate bond in the arithmetical continuum. Sometimes I think I understand the objection, but other times I lose the thread. For example, there are jumps in ordinary quantum reality. Moreover, subatomic particles are intimately connected while remaining individual elements. Quarks are entities that always occur in groups and cannot be observed confined in solitude. Such physically structured unities, given outward expression through theoretical relations, remain elemental even if in a separate sense. The continuous form progression provides an arithmetical example.

A second resolution to the antinomy is that there is no single such property as continuity. This is a cluster concept. What we have learned about explication might make us wary of the idea of a cluster concept. Our problem is that there are mutually incompatible theories of the continuum. The proposed solution is that the cluster concept of them (of course, this would not be like theories conjoined by conjunction) is sufficiently rigorous.

In that case, however, how is one to adjudicate the standoff between points and stretches? Both sides agree that there is a question, but one answer involves positing points that do not stretch, while the other involves regions that stretch. This discussion over whether or not the continuum has elements need not presuppose that the same objects are under discussion. There is an antinomy here, after all, if one decides that the continuum both has and does not have elements. While explication eliminates problems, a cluster concept would seem to compound the problem. One explication is all that was needed before, but now there is an array of issues. We seem to have taken a step in the wrong direction. "All are right" would seem to mean that any one of them will do the job. One wants to clean the sink. Both ammonia and bleach do the job, either one will do, but the sink will not get cleaned by using them both at once. "But does it not tell us something important that all these various reductions are possible?" Yes, of course it does, it tells us that we can see the continuum in all these aspects.

"They are all part of it." Well, yes and no. These newly found parts are neither points nor regions. One of the major reasons to adopt an analysis based on stretches is in order that the parts are like the whole. The problem with the points-based account is that parts are not like the whole. But with an array there is no hope to recover the idea that the parts are like the whole, for one part is quite different from another. I have already discussed the issue of separability. An individual grape is cut in half in an entirely different way than it is plucked from a cluster of grapes on a vine. What becomes of the discussion of the separability of the continuum?

"Still, the continuum is the sum of these aspects." That has the unwelcome consequence that it changes over time. Once upon a time, no one had used infinitesimals for continuity, but now that is part of the current array of expert views of the continuum. It would be hard to believe that the introduction of the new result changes the continuum. It is our current understanding of the continuum that changes. By that I mean that we see the continuum in this array of aspects. This suggests that a distinction be drawn between our current array of conceptions of the continuum and the concept of the continuum.

The difference is that the conceptions are the various interpretations, arrayed together in, for example, a canonical formulation; the concept is the meaning. There is a name in common between these accounts. If one threads these together, the new thread is not itself based on points that can be coalesced into regions, nor is it based on regions that are in any way like the whole. Moreover, the extended concept cannot mean that the continuum both has elements and does not have elements. Such a concept would not have workable boundaries, and so would not be useful in formal investigations.

The extended concept faces relational problems. It includes different relational systems that are incompatible, and in conflict. Worse, since no single relational system governs the whole, the parts cannot be like the whole. In addition to this qualitative problem, the extended concept faces a quantitative problem. It includes different incompatible assumptions about infinity. No one of these governs the whole. A hierarchy of these would suffice, however, which leads us back to the theory of sets.

All of these objections can no doubt be set aside, if, for example, the cluster is thought of as a new conception that presents an ordered array of conceptions. This conception would be far different from any one of them, requiring stretches of different kinds. In a baseball game, each inning stretches across time and yet three outs curtail it. In addition to innings there is the stretch in the middle of the seventh inning, the seventh inning stretch, for example. The analogy of the baseball game holds together incompatible attributes by not attributing them to the same underlying entity. Some group of these would suffice for a cluster concept. But here we have a single conception, not a plurality, after all, and that grouped conception would rely on regions, not points.

5. Recapitulation

All explications are artificial, to one degree or another. To be artificial, then, should not count against any logical construction of the continuum. But then artificiality cannot count against a point-based view with constructed stretches. Would M. Poincaré disagree? Who knows. Moreover, a theory of regions still recovers points built out of the constructed stretch. Many more natural presentations of the continuum are highly elaborate artifices.

When I watch a 35mm movie, I perceive no break in time or space, even though the celluloid film consists of discrete frames projected on a screen at 24 frames per second. There is the uninterrupted perception of elements. Although this does not count as evidence that continua are elements, if one insists on stretches, a stopgap measure appears. Perhaps the continuum has elements, but we do not perceive them. The point-based view of the real numbers is a genuine continuity, as we have seen. The continuity of nature would be our perception of such. An account based on stretches or other regions, then, would be said to capture a perception of the continuum, with different accounts capturing different facets of the perception of the continuum. But such an account would be of an aspect of our perception of the continuum, and not of the continuum itself. This account would not need to rely on my perception smudging the reality, for the perception picks up the characteristic pattern that parts of continua are continua.

A discrete theoretical framework that permits explications to be either elemental or not, without forcing or prejudicing a decision between these options, best matches the current expert considered convictions about continuity. An example of such a framework is the canonical first order logic with the standard theory of sets. Sets have the advantage of being themselves discrete. The deductive requirements are clearly gauged. Axioms are added as hypotheses. The standard hypothetical structure posits points, while another approach posits stretches. Both are right, but neither is inclusively comprehensive. In other words, the current status quo in the scientific community has been perceived as cleaving to a monolithic view of continuity, but it turns out to be more nuanced than that. Without a single comprehensive theory that captures the diversity of phenomena connected to the continuum, we make do with what we have. Research can be conducted with an elemental conception like points on the real number line, or not. In this sense, the scientific community has achieved a delicate balance if not a resolution, a makeshift but working way around the antinomy.

That continuity requires an infinite series is a mark of the phenomena under our consideration. The object of investigation somehow escapes the methods of ordinary human experience. This is puzzling, because it is suggestive of the idea that mathematics is one step removed from reality. Mathematical and physical reality do not seem to be completely on par with each other. Physical objects, for example, are not separable from physical processes. Since they are formal, and a step removed from experience, mathematics and logic seem to have a different character from natural science. Is there a way to separate out formal aspects of the continuum from the realm of natural science?

The canonical notation of first order logic with the axioms of set theory seems to settle the matter, but not because elements are like points. They are both *singular* but points are geometric, aren't they? It settles the matter because it shows that one can begin with sets and membership and build up any of the reductive accounts of the continuum, both stretches and points.

"Back to the shot and the jelly? And what of the division of the continuum? That whole issue? Are you yourself standing on shot or jelly?"

When I am thinking about the continuum, and its division, I think immediately of Dedekind and his cut, a paradigm of theoretical analysis. The *Schnitt* captures the jelly separation of a division of the continuum into two, the stretch to the left taken as a set becomes the means to specify that real number that marks the division and fills the gap at the cut. That is a deep cut. It fuses a stretch of the number line into a real number.

6. In the Middle of Things

The twentieth century began with a bold attempt to produce a purely logical analysis of mathematics. The reduction to pure logic failed, but logical construction in the form of explication persists. A philosophical explication of a concept presents a new aspect of the subject of investigation. The explication of mathematical concepts is sometimes associated with the analytic, for it seems plausible that mathematics has a different character from natural science; however, although efforts were made in the twentieth century to distinguish these realms, it has been difficult to forge a principled distinction from the perspective of the philosophy of mathematics. More generally, if our methodology is to begin in the middle of things rather than from a Cartesian standpoint of first philosophy, then Quine's rejection of the analytic-synthetic distinction can be seen as a transformation of the view that there is no ground for philosophy separate from that of natural science.

It is desirable to have the method of science and philosophy fit the object of investigation. But when it comes to the continuum, which involves processes that are continuous in nature, the fit depends on rules, procedures, and step-by-step constructions that seem to be at a distance from natural continuity. This puzzling aspect of the investigation of the continuum persists. Entirely different explications of continuity are available. Some resolve it into discrete points, and others resolve it into continuous regions. This does not seem to be a dispute that is merely verbal, but instead these approaches identify really divergent aspects of the continuum.

CHAPTER 7

The Linguistic Analysis of Sets

1. "Two Waves"

I have left a stone unturned, the linguistic analysis of sets and the continuum. Perhaps instead of sets being real, they are merely an ornament of our language. This skeptical argument is the topic of this chapter.

The influence of the linguistic analysis on Russell was dramatic. Once Russell made the shift to a linguistic analysis, he never went back. When did this shift happen? Russell says that it happened in two waves. On his own philosophical development Bertrand Russell will recall the impact created by Ludwig Wittgenstein as coming in "two waves," before the First World War and then again just after the war, when Wittgenstein sends him the manuscript of the *Tractatus*.* An acknowledgement in the preface to the *Tractatus* indicates that the impact was mutual. But Russell will also say, "I do not feel sure that, either then or later, the views which I believed myself to have derived from him were in fact his views." I call this Russell's penetrating recollection.

The August 1919 correspondence between Russell and Wittgenstein verifies Russell's penetrating recollection. These two letters reveal substantially different perspectives on numbers, the ancestral relation, and classes. In particular, for Russell '$1 + 1 = 2$' is true under certain logical conditions; for Wittgenstein it cannot be, strictly speaking, either true or false, although it does maintain the appearance of truth.

This influence will prove decisive. When Wittgenstein comes back to Cambridge in 1929, and restarts his work there, Russell is not really surprised which way he is going. Moore follows the development closely, as indicated by his excellent notes taken 1932–1934. Russell and Moore kept up with the later developments, which had continued on a trajectory from what they had seen years earlier. In summary, it is the *Tractatus*, not the later ordinary language philosophy, that convinced Russell that "the propositions of logic and mathematics are purely linguistic, and

* [**103**], p. 83.

that they are concerned with syntax."* Moreover, Russell himself attests to this.

We investigate that influence in this chapter.

2. Pure Form

Wittgenstein arrived in Cambridge before the publication of the second volume of the *Principia Mathematica.*† He seems to have approved of the account of arithmetic in the *Principles,* but to have been less happy with how it turned out in the *Principia.* In the second volume Whitehead and Russell deduce that $2+2=4$. For Russell this is part of logic and had been since he wrote the *Principles,* which we have discussed above. Logicism is wide open, suggestive of not only a methodology in the foundations of mathematics but also a variety of philosophical positions.

At the minimum it is the position that elementary arithmetic is logic, that logic secures the truth of the propositions of arithmetic. Elementary arithmetic is the theory of the natural numbers, $0, 1, 2, \ldots$, and the operations of addition and multiplication on them. This minimal case for Russell's logicism is an alternative reading, but it is non-limitative and coordinate with Russell's work in mathematical logic. The regular formulation is that for Russell pure mathematics is part of logic. Predicativism characterizes the foundational approach of the *Principia.* This definition makes Russell something of a logicist. Wittgenstein was himself close to logicism. Indeed, he qualified as one when he wrote about a decision procedure for mathematical logic in letters dated November and December 1913.‡

But later Russell and Wittgenstein's paths diverged: beginning in 1914 Russell advanced a new position, logical atomism, yet, in regard to logicism his basic view was unchanged; however, when Wittgenstein wrote Russell in March 1919 to announce the completion of his book, he said that "it upsets all our theory of truth, of classes, of numbers and all the rest."§ He told Russell that "it's all as clear as crystal." That summer Russell read Wittgenstein's book. On 28 June 1919 John Maynard Keynes writes Wittgenstein from Cambridge: "Your book has safely reached me and I am forwarding it immediately to Russell. As soon as peace is safely signed, is there any good likelihood of our seeing you in England?" In December Russell will meet Wittgenstein in the Hague to discuss the book; but the exchange of letters that precedes this meeting indicates a real divergence over philosophy.

---

* [**96**], p. 306. † [**122**]. ‡ [**125**], pp. 56–61. § [**125**], p. 111.

3. Logicism

The thrust of the two key August letters is no surprise, for they concern the tautology. But Wittgenstein seems to want to convince Russell that the tautology notion is not introduced as a clarificatory philosophical concept, as Russell might have thought, but rather as an operational tool. The philosophical clarification, on the other hand, is the show/say distinction.

On the thirteenth Russell writes from Wales: "I am convinced you are right in your main contention, that logical props are tautologies, which are not true in the sense that substantial props are true." But on the nineteenth Wittgenstein writes back from Cassino:

> Now I'm afraid you haven't really got hold of my main contention, to which the whole business of logical props is only a corollary. The main point is the theory of what can be expressed (*gesagt*) by props, i.e. by language (and, which comes to the same, what can be thought) and what cannot be expressed by props, but only shown (*gezeigt*); which, I believe, is the cardinal problem of philosophy.*

What can be expressed by language, the thoughts, are for Wittgenstein expressed by propositions. Russell had wanted further specifics: "I do not understand why you are content with a purely ordinal theory of number, nor why you use for the purpose an ancestral relation, when you object to ancestral relations. This part of your work I want further explained. Also you do not state your reasons against classes." In these specifics lie several points of contrast between disparate pictures of mathematical logic.

The manner in which Russell continues suggests that the symbolism itself is at stake:

> The theory of types, in my view, is a theory of correct symbolism: (a) a simple symbol must not be used to express anything complex; (b) more generally, a symbol must have the same structure as its meaning.†

In his reply from Italy on the nineteenth Wittgenstein explicitly rejects Russell's description of the symbolic notation: "That's exactly what one can't say. You cannot prescribe to a symbol what it may be used to express." Some details are now required to fill in this simple pattern of difference.

During the first decade of the twentieth century Russell experimented with different theories to thwart the paradoxes as he

---

* [**125**], pp. 121–126. † Ibid.

endeavored to reduce the concepts of pure mathematics to logic. In particular, among these is the paradox we have seen involving the class that consists of those classes that are not members of themselves. He wrote to Frege about the paradox in 1902, as we have seen. He then commenced his experimental project. His proposed solutions include the Zigzag Theory, the Theory of Limitation of Size, and the No Classes Theory.*

In the Zigzag theory classes are set by certain special propositional functions. The Theory of Limitation of Size restricts certain classes in a way first proposed by Cantor in his 28 July 1899 letter to Dedekind.† This solution is generally accepted today, embodied in the formal restrictions provided by the axioms of ZFC.

Russell says:

> We need further axioms before we can tell where the series [of ordinals] begins to be illegitimate. For, in order that an ordinal α may be legitimate, it is necessary that the propositional function 'β and γ are ordinal numbers less than α, and β less than γ' should be predicative.‡

On the No Classes Theory propositional functions appear through values but never as arguments. He finally settled on the ramified theory of types, then adjusted the system for each edition of the *Principia*.§ This theory organizes propositions and propositional functions into a hierarchy such that a function is typed over the type of its arguments, like in the extensional case. But for Wittgenstein functions are unlike operations: functions have the accidental feature that they cannot take themselves as arguments. They lack generality.

On the thirteenth of August Russell writes: "It is necessary also to be given the prop that all elementary props are given." To this Wittgenstein responds:

> This is not necessary, because it is even impossible. There is no such prop! That all elementary props are given is SHOWN by there being none having an elementary sense which is not given.¶

Wittgenstein uses strong language, but it is no wonder Russell asks, "How about generality?" Russell wants to know about the general form of the truth-functional operation N.

Wittgenstein writes back:

---

* [**106**]; cf. [**41**], p. 343f. † [**117**], pp. 113–117. ‡ [**106**], p. 44. § [**102**]; [**84**], pp. 70f. and 77ff.; [**86**], III.xi,34; [**127**], 3.33–3.341. ¶ [**125**]

> You are quite right in saying that "N($\bar{\xi}$)" may also be made to mean "$\sim p \vee \sim q \vee \sim r \vee \sim \ldots$." But this doesn't matter! I suppose you don't understand the notation of "$\bar{\xi}$." It does not mean "for all values of $\xi \ldots$."*

His analysis is to finite sequences presented through a notational iteration. Russell and Wittgenstein are at odds over the status of logic and mathematics. Russell's hypothetical approach is the "scientific method," for which see his remarkable book *Our Knowledge of the External World.* For Russell the main justification for the adoption of axioms in mathematical logic is inductive. Whitehead and Russell in the second paragraph of the preface to both the first and the second editions of the *Principia Mathematica* say that "the chief reason in favor of any theory on the principles of mathematics must always be inductive, i.e. it must lie in the fact that the theory in question enables us to deduce ordinary mathematics." For example, for Whitehead and Russell the Axiom of Infinity is a hypothesis adduced when relevant, since they could not establish it a priori as unconditional but required its result: "It seems plain that there is nothing in logic to necessitate its truth or falsehood, and that it can only be legitimately believed or disbelieved on empirical grounds."†

Wittgenstein rejects the possibility that there is any friction between Russell's Axiom of Infinity and what physics determines about the world. But Russell had recognized the incongruity.‡ Strictly, the worry can be mooted. The Axiom of Infinity is not in play in the context of this discussion insofar as it is not required in the system of the *Principia* to secure elementary arithmetic.§

Furthermore, since there is no obvious inductive reason for switching to the Sheffer stroke in the second edition, the motive guiding that decision being "to make the number of undefined ideas as small as possible," the chief reason cannot be the only reason.¶ For Wittgenstein, however, regularity is unique to logic: outside logic there is no necessity, no certainty. One might describe a "logocentric predicament" faced by Whitehead and Russell. To "give an account of logic, we must presuppose and employ logic."

Whitehead and Russell write:

> It is to some extent optional what ideas we take as undefined in mathematics. . . . We know no way of proving that such and such a system of undefined ideas contains as few as will give such

---

* [**125**]; cf. [**42**]. † [**105**], p. 183. ‡ [**127**], 5.535. For Wittgenstein on the Axiom of Infinity in 1919: [**103**], p. 87. Cf. [**1**], pp. 146–149; [**7**], p. 296. § [**75**]; [**131**]. ¶ [**123**], p. 91; cf. [**127**], 6.37f. and 6.3751; cf. [**113**], p. 228.

> and such results. Hence we can only say that such and such ideas are undefined in such and such a system, not that they are indefinable.*

But that does not prevent Russell from metamathematical reasoning.

One example occurs in "An Essay on the Foundations of Geometry."† Projective methods induce a correlation between Euclidean and non-Euclidean geometries, Russell said, that

> removes once for all, the possibility of a lurking contradiction in Metageometry, since to a proposition in one, corresponds one and only one proposition in the other, and contradictory results in the one system, therefore, would correspond to contradictory results in the other. Hence Metageometry cannot lead to contradictions, unless euclidean geometry, in the same moment, leads to corresponding contradictions.

Here already in 1897 Russell is talking about relative consistency.

Again, Kripke says that in the *Principia* Whitehead and Russell do in a sense establish metatheorems. For instance, concerning sentential logic, it states that under the hypothesis that there exists a unique x singled out by a description ϕ, any two disjunctions that differ only as to the scope of ϕ are indistinguishable with respect to truth.‡

Kripke cites an introduction to Gödel's Completeness Theorem by Dreben and van Heijenoort, who write:

> For Frege, and then for Russell and Whitehead, logic was universal: within each explicit formulation of logic all deductive reasoning, including all of classical analysis and much of Cantorian set theory, was to be formalized. Hence not only was pure quantification theory never at the center of their attention, but metasystematic questions as such, for example the question of completeness, could not be meaningfully raised. We can give different formulations of logic, formulations that differ with respect to what logical constants are taken as primitive or what formulas are taken as formal axioms, but we have no vantage point from which we can survey a given formalism as a whole, let alone look at logic whole.§

* [**123**], p. 91. † [**95**], sec. 39; cf. sec. 11; cf. [**118**], pp. 12f. ‡ [**62**], p. 1013f.
§ [**43**], p. 44.

The sentence beginning with 'Hence' suggests that they have in mind questions like completeness. No matter what they intended, clearly this sentence cannot truthfully mean that all metasystematic questions did not come up. In fine, they create a context for Gödel's breakthrough Completeness Theorem by way of historical contrast but may have overemphasized that contrast.

Russell's ideas about syntax and the precise specification of the formalism are unlike many contemporary ones. In 1931 Gödel showed that the assumption that the system of the *Principia* is ω-consistent implies there is a formal (and arithmetical) proposition $Av(v)$ constructed in the formal system of the *Principia* such that neither it nor its negation is provable in the system of the *Principia*, provided that the system is consistent. $Av(v)$ says that there is no Gödel number of a proof of $Av(v)$ in the formal system of the Principia.* Russell lives in a time before Gödel isolated formal truths that do not have proofs.

Overall, Russell responds to the paradoxes within his hypothetical framework with logic given at the outset, but the universalism is not completely evident. Having considered the logical symbolism, I turn to number. For Russell numbers can be translated into logic. Russell recognizes that Wittgenstein has a different approach: "No numbers in logic. Why?" Russell's question is perspicacious: on Wittgenstein's definition the natural numbers are exponents of an operation.†

3.1. Exercises.

(1) Did Russell work at all on metamathematics?
(2) Did our definition of the counter sets use an operation?

4. Arithmetic

In his fifth lecture on the philosophy of logical atomism delivered in London in 1918, Russell maintained that pure mathematics falls entirely within the realm of logic, for all of the propositions of pure mathematics "cannot only be expressed in logical terms but can also be deduced from the premises of logic, and therefore they are logical propositions."‡ Yet, a philosophical problem about the status of these propositions had recently arisen, he said. The problem is finding an accurate characterization of the propositions of mathematics that "can be expressed in the language of pure variables but are not propositions of logic" because they "cannot be proved from logic." Russell considered as an example the proposition 'There is at least one thing in the world.' "That," he continued in his fifth lecture, "is a proposition that you can

---

* [**46**]. † [**127**], 6.021. ‡ [**109**].

express in logical terms. It will mean, if you like, that the propositional function '$x = x$' is a possible one." The problem with '$x = x$' is that "you cannot know from logic whether it is true or false. So far as you do know it, you know it empirically, because there might happen not to be a universe, and then it would not be true." This new aspect might have something to do with the general form of the proposition.

He said: "Everything that is a proposition of logic has got to be in some sense or other like a tautology. It has got to be something that has some peculiar quality, which I do not know how to define, that belongs to logical propositions and not to others." This quality must be a priori.

But Russell's logicism is not necessarily shared by Wittgenstein. Whereas for Russell the propositions of mathematics have a determinate truth value like the propositions of natural science, albeit sometimes with varying degrees of certainty, for Wittgenstein the propositions of mathematics are "apparent propositions" only, unlike the propositions of natural science, which have a sense, yet also unlike the tautologies. An example of a tautology is 'either it is raining or it is not raining.' Wittgenstein says that knowing that tells me nothing about the weather.* Tautologies and contradictions are limit cases that default to truth and falsity but, strictly speaking, do not convey a sense: they are "*sinnlos*," making tautologies substantively different from other propositions. A tautology has no content. It is a pure form of presentation that the propositions of mathematical logic appear to share, but only under false pretenses.

In the *Tractatus* the propositions of elementary arithmetic do appear to be like the tautologies. This is largely because the cast of Wittgenstein's operation is much like Church's lambda calculus, for example. Unlike the tautologies, however, for Wittgenstein the elementary propositions are apparent propositions only, so there is no real basis in logic. They are apparent because they are not propositions of logic at all. They only look that way. For Wittgenstein the equation '$2 + 2 = 4$' is not a tautology. He writes about this in a typescript commonly identified as TS 206.

One of our great authorities on Wittgenstein and mathematics connects the elimination of identity to the superfluity of the theory of classes.† Wittgenstein will tell Ramsey that identity is not tautological. Identities cannot be true, even in a derivative sense. The arithmetical proposition is not based on the iteration: identity is not generated by an operation.

The analogy between equation and tautology is the key. For Russell that is a fair analogy, while for Wittgenstein it is not. For

---

* [**127**], 4.46f. and 4.462f. † [**70**], p. 354; [**52**], p. 228.

Wittgenstein tautologies are results of an operation, while identities require proof.* The ancestral relation for Frege and Russell secures the logical status of mathematical induction. In the *Tractatus* the formal series stands in for the ancestral relation: "We can determine the general term of the formal series by giving its first term and the general form of the operation, which generates the following term out of the preceding proposition." The consequence is that mathematical induction is not logic for Wittgenstein; however, tautologies are not "unsinnig."† Wittgenstein says the proper analogy is between tautology and number, because tautologies are part of the symbolism much like in mathematics '0' is part of the symbolism. Equations do not share this characteristic.

5. Classes

Having considered the status of the arithmetical equations, I turn to classes. Russell had reduced the concept of cardinal number to the concept of class in section 111 of his *Principles*. As we have said, although for us class has a precise meaning as distinct from set, by "class" Russell means any abstract collection, including sets. Russell and Whitehead discuss small classes in the second volume of the *Principia*.‡ The number of a class α is the class of all classes similar to α: $\{\beta : \beta \sim \alpha\}$. A number, then, is anything which is the (cardinal) number of some class. Russell says that numbers are classes of classes. On this definition a number is a property of a concept and is part of logic.§ Russell continues to call this definition in later writings.

In 1918, in his eighth lecture on logical atomism, he said that classes are "logical fictions."¶ Since numbers are classes of classes, he concluded that numbers are logical fictions. A number he says in his sixth lecture is a logical fiction from the perspective that a number is an "incomplete symbol." But this is nothing new, it is the case already in the Principia. He said of classes: "A statement which appears to be about a class will only be significant if it is capable of translation into a form in which no mention is made of a class." He said: "This places a limitation upon the ways in which what are nominally, though not really, names for classes can occur significantly: a sentence or set of symbols in which such pseudo-names occur in wrong ways is not false, but strictly devoid of meaning."

It is plausible that Wittgenstein's formal variable need not involve the postulation of a class. Russell himself had adopted the Axiom of Reducibility to preserve intensions, but Wittgenstein

---

* [**127**], 6.241, 4.1273; [**38**], sec. 26. † [**127**], 4.4611. ‡ [**122**], p. 101.
§ Cf. [**39**]; [**40**]; [**111**]; [**99**], p. 327. ¶ [**109**].

rejects that option.* Russell is puzzled by Wittgenstein's remark that the theory of classes is superfluous in mathematics:

> You deny classes, so cardinals collapse. What happens to $\aleph_0$? If you said classes were superfluous in logic I would imagine that I understood you, by supposing a distinction between logic and mathematics; but when you say they are unnecessary in mathematics I am puzzled. E.g. something true is expressed by $\mathrm{Nc}`\mathrm{Cl}`\alpha = 2^{\mathrm{Nc}`\alpha}$. How do you restate this prop?

Wittgenstein does not answer Russell's question in 1919.

One answer might be that the proposition expresses a class variable for the natural numbers, that '0' or '0'x' marks a "formal property."† A natural number for Wittgenstein is a property of an operation. It stands for a mathematical concept. This could be secured by way of the notational iteration. There is nothing in the symbolism that bars this answer, but what supports the notational series from collapsing? Wittgenstein's claim about the superfluity of class theory at the minimum would seem to require that his operation be monotone and continuous, and it is Russell in his introduction to the *Tractatus* who will first read the notational procedure as an infinitary operation.‡

Once the iteration is read as providing for infinite sums or products, the truth of the arithmetical equations is secured. This might require an appeal to infinitary reasoning to bolster the claim about the superfluity of class theory. This transcendence means the arithmetical equations would come out true. Yet, one can deny they do.§ The noteworthy lineage for this reading traces back through to Russell himself. But the sequences associated with infinitary operations are not the sequences Wittgenstein had in mind, he will later write.¶ The limit case is a continuation of a different sort, which he did not seem to realize back then.

Although many reconstructions are possible, Russell's unanswered question hits its mark. The *Tractatus* does not access the transfinite. One must assume the Axiom of Infinity to get there from here. By "to get there from here" I mean that Wittgenstein's model would work were it populated with an infinite number of constant symbols or names. But he never considered that and had not distinguished carefully the finite and infinite cases.

---

* [**125**], pp. 35 and 53; [**127**], 6.1232f. † [**69**], 111f. ‡ [**74**], p. 229. § [**68**], p. 96; [**69**], p. 114. ¶ [**126**], p. 268.

6. Historical Assessment

6.1. If Russell and Wittgenstein once held similar views on mathematical logic, their correspondence during the summer of 1919 shows substantially different perspectives on numbers, truth, and classes. Russell is a logicist more like Frege; for Wittgenstein a proposition of elementary arithmetic is not really a proposition at all but is only in the guise of a proposition, so its truth is not really a possibility but is rather something of a category error.

6.2. Their correspondence is not idiosyncratic, since the pattern of differences it exhibits repeats in other work. Both write about the laws of physics, but Russell conceives of this as part of the larger project and within the logical framework. The theory of probability girds the Law of Induction, which is a priori, if true, Russell says.* Wittgenstein rejects the possibility that Russell's Law of Induction is logic because it has a sense, and for the same reason it cannot be a priori.

Russell is not really attempting to incorporate Wittgenstein's view of logic as pure form without content. He is more engaged in his own overarching scientific enterprise through mathematical logic and toward the analysis of matter. Russell is a realist for whom what is described is to be independent of the description. But Wittgenstein is an idealist about logic. He says that "logic is transcendental."†

6.3. Is Wittgenstein still a logicist when he writes the *Tractatus*? Sure, let us say we call it *transcendental logicism*. But this is in the derivative sense of giving a logical schematism or scaffolding rather than a reduction, per se, for it is a condition. After all, he claims that logic is a priori and that we cannot think illogically. That qualifies as "enough said," although clearly he is not a logicist like Frege or Russell. As we have seen, his approach to truth is quite different.

Is Wittgenstein a logical atomist, like some have insisted? I do not know. I thought that logical atomism was Russell's philosophy from 1914 to 1918.‡ I thought the logical atomism arose in response to the *first wave* of Wittgenstein's influence on Russell. But we have not discussed that here.

Later on, Wittgenstein introduces the language game, which can be seen as a sort of cultural manifestation of concept formation. He thereby gains a certain universality across cultures. My friend Joseph X. Donovan says Wittgenstein is more universal than Russell in this light. We discussed universalism above as a kind of position where there is no metareasoning. Wittgenstein

---

* [**107**], p. 225; cf. [**105**]; cf. [**127**], 6.31. † [**127**], 6.13; cf. [**71**]. ‡ [**109**].

rejects metamathematics and his language game can be seen as facilitating such reasoning without *jumping up a level.*

What about us? We live after the proof of the incompleteness of any formal system, which means we are not logicists, in general, although there have been some recent interesting attempts to salvage this great philosophy. Also, we rely on metareasoning in mathematics and across the natural sciences. We cannot prove the consistency of a formal system from within it.

6.4. Try as he might, Russell is always so close to the rationalist roots, and to Leibniz. He cannot get out of that. What many do not realize is how deeply finitistic Leibniz was and rationalism often is. The rational realm augments Russell's world view sufficiently such that infinity should not be needed, but he cannot prove infinity. Since Quine is not a rationalist, he is always touching ground, returning to the world of friction and motion, but sometimes less so for Russell.

Perhaps it is this that makes him liable to Wittgenstein's linguistic analysis. If one comes from a less rational and more empirical basis in philosophy, if one begins with questions in science that move to the foundations of science and then to sets, one sees the power of the linguistic analysis, and also its transparency.

A linguistic analysis, one must remember, wants not to be deep but shallow. And so it is. The hope of Wittgensteinians is that there is some special insight to be found in the concept formation of Wittgenstein's later philosophy that is not merely linguistic.

CHAPTER 8

Worldmaking and the Continuum

There was a fitness of things in the whole
that carried conviction with it.
Dickens, *Bleak House*

1. Summation of the Last Several Chapters

1.1. As even the casual reader will have guessed, I do not agree with Wittgenstein that set theory is superfluous. Of course, one should add that Wittgenstein, in the long run, was doing something of a cultural analysis. He was not doing theoretical philosophy but wanted to investigate concept formation, to describe concept formation without explaining it. One quickly recognizes that concept formation, how that comes about in each of us, is not uniform, whether the concept is, or not.

One great example of this is the Galilean infinite. We did a version of this with sheep in chapter three. Galileo Galilei compared the set of squares

$$1, 4, 9, \ldots$$

or

$$1^2, 2^2, 3^2, \ldots$$

with the set of positive integers. What happens when one catches the distinction between the infinite situation, in which proper subsets β of an infinite set α are equinumerous with α, and the finite situation, in which all proper subsets β of a set α map bijectively into but not onto α? A light goes on.

But how that light goes on and all that is involved with that, the individual process seems quite open-ended. When I teach the Galilean infinite in class, so many students resonate with it directly. And when I reflect with them about it thereafter, each relates uniquely. I find it a bit startling how variegated the responses are over many years of teaching this material. This is evidence to me of different ways of knowing the same thing. But I do not think that the subjectivity of those different paths can be made objective.

Conceptual boundaries are clear cut for Russell; they are not for Wittgenstein in his later work. There is much disagreement over whether or not Wittgenstein believed that hazy conceptual boundaries can be distilled and clarified in particular formal cases for the purposes of theory. I have argued that proof in mathematics is the main condition for conceptual clarity in his *Remarks on the Foundations of Mathematics*. If so, he did believe one can formalize games in particular narrow cases. But then he is not a genuine skeptic. There are many ways to understand what a linguistic analysis comes to, not all equally convincing, perhaps. Russell himself, and many others, came to think that a linguistic analysis reveals mathematics to be empty of any content beyond formal content.

But we have reason to think otherwise. We have seen there are no resources within the analytic itself that are somehow isolatable as sufficient for the foundations that sets provide. We have seen that sets can be considered a basement for elaborate constructions that no longer need to refer to sets. Of course, in the sense that one can do category theory instead of set theory, set theory is superfluous. But category theory is confirmed as a foundational theory because it can exactly match the results of set theory. We can avoid sets by doing category theory, but then we have categories. Since mathematics can be founded on sets or on categories, neither is innocent. They are two paths to the same end.

And we saw that for example different definitions of the ordered pair will all do the trick. One definition makes the others superfluous. If so, the superfluity is just an empty remark, like a comment made in spite, and not a serious philosophical consideration. Tit for tat, one might respond, "No, category theory is superfluous." But that is not an argument. Category theory and set theory are dual notions. It does not seem that Wittgenstein was saying that set theory is superfluous because another theory can fill the bill. He was saying we do not owe any money on this bill.

1.2. But then again consider for example the cross-pollination between the foundations of mathematics and so many branches of science, like for example between sets and quanta. Quantum theory and quantum field theory require Hilbert spaces. Hilbert spaces are part of functional analysis. Functional analysis generalizes out of n-dimensional space to these Hilbert spaces of infinite dimensions. The n-dimensional space is based in linear algebra and the theory of abstract spaces, like non-Euclidean geometries. Space-time is an example of an abstract space. Space-time is the form of the material world. Its curvature, due to gravity, is studied in physics by relativity.

Much of mathematical analysis was organized from the end of the eighteenth to the beginning of the nineteenth centuries in what we now call the theory of functions of a real variable. The Spectral Theorem itself, which lies deep in the foundations of quantum reality, falls under the theory of functions of a real variable.

Now, according to an expert, Sergey Borisovich Stečkin:

> The foundation on which this theory of functions of a real variable is built is the *theory of sets.**

I suggest that theory of sets lies deeply in our theoretical understanding of physical reality. It is far from superfluous.

Perhaps 100 years ago, when set theory was still fairly young, the remark about superfluity would be somehow powerful. But in this day and age with the results on the independence of the Continuum Hypothesis, the work on large cardinal axioms, and much more, it looks from our perspective to have been somewhat misguided.

Sets have proved to be well-suited for foundational investigations from the bottom-up. And the foundational investigations have not been empty, after all, but have reaped benefits across the sciences. Researchers continue to rely on them.

For all I can see, they are part of mathematical reality. And when it comes to cases like the continuum, cases in which the mathematical and the physical intersect, there is no principled way for one to insist that they are not part of physical reality. Or so I have argued.

1.3. "But is this not *your* view?" But of course it is.

"Well, then, why do you keep talking about reality?" Well, I am obviously not talking about reality as I do not see it but as I see it. And my report reflects not my subjective judgment on reality but what the experts tell me. I am reporting out. But I do not think that is a sufficient answer, for I myself find finite approaches to the continuum highly persuasive and plausible. "Yes, is this not a world that we have made? And what exactly do you mean by 'reporting out'?"

My problem at present in terms of this line of argument is that I bailed on this topic many chapters ago, when I mentioned Weyl but did not give him his due. Here at long last I return to that topic. But this is not a return, it is too late for that, it is a new topic. I simply cannot go all the way back to where we were four chapters ago and restart on that and do it justice.

Instead, carpe diem. Where are we now? We began with sets, moved to the continuum, then turned to its analysis, before last chapter's work on linguistic analysis. Here in this chapter we

* [**114**], p. 4.

focus on nominalistic views. Actually, this has been a theme from the onset.

2. Catherine Elgin

An expert on world making writes:

> Rather than reflecting without distortion the way the mind-independent world is, theories refract. They provide an understanding of the world as modulated by a particular theory. Truth is defined within a theoretical framework rather than outside of it. This does not undermine objectivity, for an assertion contains a reference to the framework in terms of which its truth is claimed.*

This opens a persuasive case that, in regard to coherent representation, the arts and sciences are not too far apart with only shades of real difference between them.

There is a nuance, for this view depends on a new sense of objectivity, which Catherine Z. Elgin defends with a purely procedural and pluralistic epistemology. The result is the collapse of consistency, and even of relative consistency, across communities, a steep price that the nominalist choosing this line must be willing to pay. In this chapter we slightly attenuate this structure to incorporate the firm positions of previously unnoticed exemplars.

We might for the sake of investigation supply a constructive nominalism to replace scientific realism, to see how far that might go. Theories on this account rather "refract than reflect a mind independent world, modulating our understanding of the world. Truth is defined in the theoretical context, not outside it, but this does not damage objectivity," as "reference is framework oriented." Here we glimpse a new sense of objectivity. Central to it is "the standard of rational acceptability," namely reflective equilibrium.

It seems to me there would be an impact on the expert understanding of relative consistency, like the consistency of the Continuum Hypothesis relative to ZF, which was proved by Gödel. The idea of relative consistency is to test a new axiom. One takes a set of axioms which one presumes to be consistent, then one shows that under that assumption, the new axiom is consistent with them. Relative consistency is a huge advance in our understanding of mathematical reality through axiomatic structures. But also, without truth there is no existence in the traditional sense. To say that something is, is to say that the sentence that involves it is true.

---

* [**28**].

More prosaically, what about keeping promises? How can one keep a promise without truth? Let us consider a world that I build. Can a promise be kept in a world that I build. Yes, but it bears promise only in the world. Can a promise be kept in a world we build? No, of course not unless the promise is not a true promise. Of course, I can believe in a promise in a world we build. But I can also believe in dreams. Should we now deny belief? Belief in a promise is not enough for promising. We owe each other more than that.

Say it is our world. My question is, can we be together in a truly committed way in a world we build together? Well, can we be apart? No. What, then, can it mean to say we can be together?

3. Reflective Equilibrium

Here in this section we discuss the burdens of judgment; felicitous falsehoods are true enough. Let me introduce a standard example, the gestation of the elephant. It is an outlier that does not conform to the rule of the gestation of mammals. But we do not give up the rule! This is the problem of outliers.

In a lucid 1993 paper published in a festschrift in honor of Israel Scheffler, Elgin identified a separation between objectivity and truth. A classic question concerning objectivity concerns how we test a theory:

> We test our theories, Scheffler believes, not against some fixed stock of irrevocable intuitions, but against our initially credible judgments about the subject at hand.*

"We do the same in science. Certainty is not to be had." The lack of certainty has the outcome that "even observation sentences may be given up if they clash with commitments we are loath to reject."

But this should not launch us toward pure coherentism because "a coherent system of mutually supportive statements could be a complete fiction." Indeed, it is easy to agree with Plato that objective knowledge needs a tether. But with that comes the realization that the tether ties us to facts of the matter about our own previous best guesses:

> What Scheffler and Goodman both realize is that the tether need not connect theories directly to the world. It is enough if it connects them to our antecedent commitments about the subject.

---

* [**30**], p. 8.

> These commitments, they contend, have some measure of initial credibility. They are not certain or irrevocable. But they have some presumption in their favor, being our current best guesses as to how things stand. As far as possible, we construct our theories around them, augmenting and correcting as necessary to maximize credibility. A maximally credible theory is, in Rawls's terms, in reflective equilibrium.*

Reflective equilibrium is central to our contemporary understanding of philosophical method:

> Its elements are reasonable in light of one another, for the theory is coherent. And the theory as a whole is reasonable in light of what we already had reason to believe, for it is tethered to our initially credible judgments. There is, of course, no guarantee that such a theory is true. But by now the history of philosophy should have convinced us that there are no guarantees.†

And "since the theory is maximally credible, we have reason to endorse it in current epistemic circumstances."

It is clear from this passage that this particular version of reflective equilibrium is so minimalistic that it may genuinely be asked whether it involves any commitments. Perhaps this is a form of nominalism. But before we get to that, I am still trying to forge the connection to philosophical method that Elgin takes as central to the explication of truth.

For example, 'plus' is entrenched in a way that 'quus' cannot match, so theories employing 'quus' are less credible than those employing 'plus.' On the other hand, should new counterexamples to 'plus' emerge, the initial credibility of hypotheses framed using this term will diminish, and the presumption in its favor might shift. This nice feature provides a framework for the structure of scientific progress. What, then, is objectivity?

> Objectivity, Scheffler contends, depends on methods, not results. A discipline is objective to the extent that it subjects claims to critical scrutiny, to the extent that it allows for testing by independent and impartial criteria.
>
> Scheffler acknowledges that category schemes are human constructs, framed in light of beliefs, interests, and aspirations. And he

* [**30**]. † Ibid.

> acknowledges that observation is theory-laden, being informed by habits, expectations, and conceptual commitments. Still, he maintains, science is objective. For whether our standards are satisfied is not settled in our framing of them.*

But such objectivity hangs by a thread. It "turns on testability. And not all testable judgments pass the tests they face. Moreover, by passing the relevant tests, a judgment is merely confirmed, not proved." It turns out that an "objective judgment may well be false."

What type of procedure does one envision? Following Rawls, there are three basic types of procedure, and these are distinct.† As a model for fair distributive justice, Rawls favors a pure procedure over perfect or imperfect procedures. Whatever Rawls happens to be doing, a helpful quote us of the new approach on which epistemology might be thought of as an imperfect procedure: "To err is human; to do otherwise, inordinately difficult."‡ We saw that systematic understanding of any sort is being said to be "objectual." Felicitous falsehoods are wrapped into any overall account as they increase "tenability," Catherine Elgin argues, allowing us to cast key ideas in bold relief. Prioritization permits us to adjudicate between competing claims.

What about assessment? Instead of proceeding through each component sentence to ascertain whether or not it corresponds to a fact, systematic accounts are to be assessed holistically. The focus is on the coherence of the account, but the coordinate philosophy is not entirely coherentist, as we will see. The fabric of understanding is woven together, one might say, if it stands in reflective equilibrium, and as such replaces the foundations.

4. Social Reality and the Body of Experts

In this section we consider the case of mathematical understanding, and slightly attenuate the model developed above to incorporate the firm positions of certain significant exemplars, which have not yet been considered, having been previously unnoticed. An argument concerning the principles of mathematics can be run in parallel to the art/life argument, as we will see. It is noteworthy that the focus here is rather on the principles of mathematics than on the foundations of mathematics. Elgin seems to hold that the principles are in principle unprincipled, to give a fun slogan.

Here is the argument. Due to the fundamental instability of objectivity and the procedures of reflective equilibrium to compensate, somewhat, truth is not extraneous to theory but always

---

* Ibid. † [**89**], pp. 84ff. ‡ [**22**], p. 21, 82.

within it. This needs to mesh with the disquotational view of truth. So, we stand in this world and formulate the Tarski definition because that is what the best experts do in mathematical logic. We also regiment theory with first order logic for the same reason. We need also to introduce at a minimum computation theory, arithmetic, and analysis. All of this is completely in agreement with Elgin's new kind of nominalism. Now of course we could claim that much of this apparatus may very well be, or as she would say, is, a felicitous falsehood.

But now we are inside the model, not outside. And Elgin approves of this mode theoretically. Inside the model the experts debate the question of the extent of reality. One brilliant example is that of Solomon Feferman, who argues that predicativism is the extent of reality. Others have disagreed in either direction. Strict finitists say there is less to reality while classical Platonists claim there is more.

My point is not which side to take on this debate. My point is that many of the experts agree that there is a precise issue being debated here that itself can be mathematically formulated. Thus, to honor reflective equilibrium, at this juncture we introduce the ability of principles to derive judgments in accord with the expert community. My reflective equilibrium shifts as I incorporate the new distinction into the model. Of course, I will not change everything. I still say that for all I know this all might be a felicitous and false theory.

But since we have shown that this very theory can itself bear a principled distinction, I now am forced to acknowledge that principled distinctions are still available. Not everything shades off from fiction to fact indiscriminately in the model. So it is a false generalization to go further and insist that there are only unprincipled distinctions and no real difference between theories that gave different axiomatic strengths. No one would ever say otherwise, if they have a serious theory of reference.

Let me give a parallel example from fiction to make it plain why, as long as we are free to follow the rules of those worlds as we told to do by experts, we have enough structure to enact principled theoretical distinctions in a precise axiomatic fashion so exact that a computer can do the sorting

The view involves world making. Fact and fiction are on a continuum here. Truth really does not enter into the picture. Instead, the model features waste cases, or throwaways, like the gestation of the elephant and extends that into a holistic picture. Without appeal to truth available, and apparently no recourse to either correspondence or coherence theories of truth, we must rely on expert communities. It is at this tribunal that fact is tested.

As with virtue on the view of Aristotle, that virtue is not innate or by nature but is a state of character, so one must use exemplars for guidance as to judgment. In the case of science, it is the expert scientific community that guides judgment. One is to submit to these authorities because truth is not available as a counterweight.

That is the set up. Now we enter this constructed world. Here again there is no recourse to truth, but only to the expert community. What does the expert community of artists say about any discontinuity of fact and fiction? They accept it. This is not to say that fictions are not valuable or useful, but it is to remind us that watching a movie is not watching reality. As Samuel Beckett said, "The gap between art and life is ever-fixed."

Rather than quote a series of artists, authors, poets, and movie directors who, like Beckett, find that the texture of reality is unlike that of fiction, let us simply posit that art can be very important and very realistic without in any way being confused with real life. One is supposed to know the difference between play time and work time. So, let us build into the model enough resources to distinguish the Tarski definition of truth, which obviously will be welcome. Even if the argument that the global distinction between fiction and truth cannot be drawn at the outset from below goes forward, that should not lead us to think that the distinction cannot be drawn from within. Inside the model I set all sentences that correspond to the facts as we understand them to true, and all other sentences to false. 'Leopold Bloom is the husband of Molly, but Molly loves Blazes' I set to false because 'Molly Bloom' does not refer.

But our focus is on philosophy, and many philosophers disagree about the crossover between aesthetics and science. Causes are not reasons, to put the point in a slogan. It is interesting to note, as a complexity in an historical understanding of Elgin's thought, that a looming influence on Elgin, namely Quine, would disagree with her about this position. Quine says metaphors are false. Elgin we might say agrees with Quine about metaphors but disagrees with him that science is somehow different. But is the case that ideals function in the same way in both the sciences and the arts? Let us incorporate the experts that answer negatively into our model.

To accommodate the discontinuity, we regiment by translation into first order quantification theory plus identity. We add a minimalistic Tarski definition of truth to build a semantics. Of course, from our perspective outside this world we could add in Tarski's definition. But from inside the world we, by the very rules of the world, cannot be sure that this corresponds to the actual most minimalistic definition of truth. Why should we accept it from inside anyway? Because the exemplars tell us to do that very thing.

So, while we cannot be sure in the constructed world that our definitions are the actual ones, no matter how well they work, we can indeed distinguish felicitous falsehoods from truths in Elgin's own terms to provide for the community of expert opinion. I conclude that, however much we agree with the narrower point that truth is only the beginning, it does not just disappear, for all that.

The result of this section is that, once we've modified Elgin's model in a coherent way that her system as it stands does not rule out, we can see that we can draw a sharp distinction between the fictive and the factual right in her model. Of course, since we are in her world, we cannot know whether or not that distinction corresponds to an actual distinction. But in any case, we are now in a position to articulate in her model the traditional separation of the arts and the sciences. But it had seemed that this was a distinction that she in her metaremarks would have rejected.

5. Predicativism

A plausible account finding that the continuum is mentalistic seems to have introduced by Solomon Feferman.* What makes the account plausible, at least to me, is that the mental on this account is a creature of the body of experts working together and separately on research into the continuum. The continuum is thus a social reality.

This would seem to be an appealing picture, especially when one considers that there is objectivity about the continuum on this view, for it arises from the intersubjectivity of the social reality. The view does not, then, depend on some extraordinary power of the mind. The analogy for the indeterminacy of the continuum is with the line of succession to the presidency of the United States. That is determined out to the eighteenth place, with the vice president being first, and the secretary of Homeland Security being eighteenth. But this is a formal determination, and many different actors might play these parts. While this manuscript was being composed, for example the eighteenth place has been filled by two different secretaries, Kristi Noem and Markwayne Mullin. But, something even more significant for our work is that after that, after the eighteenth place, it is not determined in any way at all.

Feferman's view is that the continuum is like that.

Not all the points have been determined, and indeed it is not even determined that the continuum is punctiform, except insofar as there is a community standard. The continuum is determinate

* [**35**].

so far as the social reality has determined it, and otherwise it is indeterminate. Feferman draws on Weyl's tradition.* Weyl's idea was to develop a logic of relations. We saw that this was a critical step for Russell, as well, but their approaches are different. Weyl opts for constructing relations. Feferman also identifies the set notion as problematic. He commits to the actual infinite, and to its objectivity.

What concerns him are subsets. We have seen some of the power of the subset, but let us not forget the power set operation, and not merely the power set operation, but the power set operation applied to the actual infinite. So, given the subset issues, it is no surprise that mentalism maintains a far different account of functions than what we have seen thus far, for example.† Feferman is profound in his remarks on subsets.

From the philosophical perspective, one wants to respond that the labyrinth of the continuum involves the infinite and relations. Those are the two doorways, and the one doorway that one often wants to close is the actual infinite. But here is a case in which the other option is being considered. The power set operation is a powerful way of relating all the members of a set. It is like saying, here are all the options for the ways the members of this set can relate or group up. Of course, they don't have to group ($\emptyset$), they can group by themselves in singletons, or with just one other as pairs, ..., or the whole group might get together. But the infinite subsets of the power set of the natural numbers, they are something else.

Feferman quips, by way of contrast with his own view, that *Platonism is the medieval metaphysics of mathematics.* There was a tradition of nominalism in the medieval that is closer to this newer philosophy Feferman espouses than he may have realized. Moreover, the idea of the indeterminateness of the continuum traces back through the medieval period past Aristotle to the apeiron, or "to apeiron," the unbounded indeterminate background condition in Plato's *Philebus* and before that in fragments attributed to Anaximander.

These historical questions move us too far afield from the contemporary action. Maybe, however, one last minor complaint is in order, for some of the best known philosophers of the scholastic tradition were much influenced by the peripatetic school. Those standing in this tradition, while they did not reject relations, did approach relations in a guarded way, for they knew about some of the questions we have seen concerning relations. A prominent position was that the foundation of a relating reality did not have an

---

* [**36**]. † [**35**]; [**37**], p. 248.

external existence like that of Aristotle or a horse. There was fundamentally a mental component. The *fundamentum* has a mental aspect in the medieval period, arguably, in the thought of these philosophers and others, and this can be seen to be something like a precursor to Feferman's view. Interesting stuff, but we need to wrap up.

Mentalism about an ordered chain of constructed forms is the idea that the series extends out determinately as far as it has been socially determined, and is indeterminate farther out. This indeterminate itself has several possible interpretations. One idea is that the rule has not yet been carried out, but a different idea is that there is no rule to determine the next step, because any series can in retrospect be captured by some rule or other, no matter how outlandish the rule is.

One might say that mentalism is self-sustaining. A mentalist has good reason to think that the continuum is indeterminate given his view of the continuum, for a mentalistic concept that exceeds the range of the mental is indeterminate. But mentalism also rejects that there is a non-mental component to the continuum itself, which puts things on shakier ground.

L is a sort of model of what nominalism comes to in considerations of the continuum. Of course, there are forcing models of set theory that, once augmented by specially designed real numbers, reject the continuum hypothesis. But forcing models are also used in arguments for the continuum hypothesis. While there is no completely safe ground as to the size of the continuum, I remain optimistic on Cantor's original insight as to the solution. Please see Appendix **C**.

6. Flashes and Specks

Let us finish up by considering the wisdom of Walt Whitman. "Wait, a second. Walt Whitman?" Yes, please remember that we are relying on the community of experts for a reflective equilibrium in our model. I shall exploit the fact that the community of experts from the arts judge that there is no overlap between fact and fiction, in large part. I want to bring in Walt Whitman here.

> The family usages, the language, the company, the furniture, the yearning and swelling heart,
> Affection that will not be gainsay'd, the sense of what is real, the thought if after all it should prove unreal,
> The doubts of day-time and the doubts of night-time, the curious whether and how,

> Whether that which appears so is so, or is it
> all flashes and specks?
>
> "There Was a Child Went Forth"

Assume for the sake of argument that objectivity is a fata mor gana. Then how can this poem, Whitman's "There Was a Child Went Forth," have any connection with us? For there can be no such doubts.

The elimination of objectivity calms our reflections, but at the cost of the sense of the real. That is, one view would be that these objects, the furniture and the rest, are not really becoming part of the poet. But then one seems to be making a truth claim after all, namely it is true that there is no truth.

For, if not making that claim, then what claim? What about the old Protagorean paradox: One asserts (as truth) that there is no truth. Now is that true or not? If one rejects truth, how can it be? So it must be false. But then how can one assert it? For this too can only be a partial truth.

'Epistemology trumps ontology' goes too far, I would argue. For example, from within the framework of reflective equilibrium I have tried to sketch a way out, as follows. The ability to construct, the fact that we do produce human constructs, in no way affects the outcome of the ultimate questions about philosophy. This is because human constructs have to do with us and our understanding and so are not entirely limitative. This marks the source of my small disagreement with Elgin's view of objectivity.

In one prominent sense, truth is a formal predicate of formal sentences that formally correspond to facts. Truth on this view simply does not itself change things. Truth is primordial, there at the beginning, and always, but not really adding anything. But scientific methods do not preserve truth absolutely. Think of whale gestation, again.

This is not to say that the problem is that we should make our generalizations true by restricting sample size. The problem is deeper than that. What if scientific methods are not unlike artistic methods? But they are unlike artistic methods, in point of fact, or so the artists say.

Concerning an "imperfect procedural epistemology," Elgin writes:

> Methods too are revisable. The best we could do yesterday need not be the best we can do today. So imperfect procedural epistemology is

> prepared to criticize, modify, reinterpret, and—if need be—renounce constituent ends and means.*

Here she cites John Dewey's *Human Nature and Conduct.*† This is no surprise.‡

I next summarize my response to the concern, associated with an imperfect procedural epistemology, about methods being under consideration for revision.

I find it hard to disagree with the idea of epistemology as an imperfect procedure. If we agree that theories are revisable and that even methods have been, and are, revisable, then we seem to be consonant with the ordinary dictates of critical thinking, logic, and empirical and scientific reasoning. But if it is an imperfect procedure, that suggests there is something about the matter to get right, and that we do not do that in our epistemology. There is, therefore, on this account friction with something beyond the mental.

As a reminder, for a perfect epistemology we have logical inference from knowledge to truth. What is ruled out is the case that I know X, but X does not exist. Again, with propositional knowledge what cannot occur is that I know that ϕ, or I know ϕ to be true when ϕ is false. In other words, for the main criteria of a perfect epistemology knowledge of a thing implies its reality.

But then it seems like one ends up a Platonist by accident. So, this is a worry that I have about the overall project, the appeal to an imperfect procedure to understand epistemology, although I initially liked the idea. For this would seem to bite off more than it can chew, in the sense that it seems to require a buildout by way of contrast toward a reality that it might have hoped to reject.

The experts are not telling us, here is a bunch of theories, take your pick. My doctor does not tell me to take my pick. She advises me on which medicine to take. The experts advise me on which theory to select. They are making commitments in choosing theories. What point is there in saying, oh yes, that is your theory but it could be otherwise?

We already know that. Theories are revisable. And for all that, whether the procedure is imperfect, perfect, or pure, theories capture our commitments. To avoid the commitments, avoid the theory. One should not have one's cake and get to eat it, too.

---

* [**22**], p. 12. † [**19**], pp. 25–35. ‡ Works by Elgin reflected in this chapter include: [**22**], [**23**], [**24**], [**25**], [**26**], [**27**], [**29**], [**30**], [**31**], [**32**].

CHAPTER 9

The Universe of Sets

1. Review

There had been longstanding interest in the principles of mathematics and in the continuum, but Cantor was the first to measure the continuum. He did this with his theory of sets. So that seemed to me a good place to start this book.

Actually, there is a tripartite structure to this book. The first part dealt with sets, the building blocks. The second part used those building blocks in the analysis of the continuum. It turned out that the analysis of the continuum advanced in the second part has led to questions deeper questions about reality and the nature of analysis. The third part of the book considered, and reflected upon, the key debates.

There is only one continuum, the problem being to come to understand it and to measure it. But I did not argue for this by fiat. I showed that there are many ways of knowing the continuum, many analyses. My maneuver was to begin with set theory itself. Any reduction of the continuum involves sets, unless it involves categories. So, I circumvented the whole argument about a plurality of continuums by redirecting to sets. The context for discussions of the continuum is a model of the universe of sets.

We saw in the second part that powerful analyses on both sides of the dispute resolve the continuum to sets. There are many ways of measuring and knowing and working with the continuum. But, if there were more than one continuum, it would have different sizes, one might be $\aleph_2$ and another $\aleph_1$. But then those two continuums would not be identical. We have already seen how to work around this. One can get around any concerns by treating the cardinality of the continuum as a hypothetical. One might build more powerful models.

We saw in the third part that how we treat the analysis of the continuum can be generalized to other realms of investigation. Work in mathematics, the sciences, and the arts all involves the explanation of phenomena by replacing the poorly understood phenomena with an explication in new terms. The explication involves an appeal to other objects in a new theoretical structure. We understand that appeal through the canonical notation.

These natural and logical constructions represent a step away from the original entity, which they ultimately clarify. Sometimes it is useful to use an analog measure, and sometimes it useful to use a digital measure. But, once one measurement has been made, the next one will conform to it, even if the device is switched from analog to digital, or else there is something wrong with the experiment.

Given the many ways of thinking about so many accounts of the continuum, some of them extremely persuasive, the intent of this book has been to rediscover some measure of the continuum. But so far we have not clearly stated how to measure it decisively. Once we have accomplished this, our work here will be done.

The goal, this is both Cantor's goal and ours, is to measure the continuum. So, we reformulated the measurement problem as a cardinality question.

How many real numbers are there on the number line? This is Cantor's Continuum Problem. It asks about $\mathfrak{c}$, the cardinality of the continuum

$$\mathfrak{c} = |\mathbb{R}|.$$

The Continuum Problem is Hilbert's first problem, by the way.* We introduced a theory to explicate the problem of the continuum as the question of how many subsets the counter sets have.

Cantor faced the problem and proposed a solution, the Continuum Hypothesis. Cantor's Continuum Hypothesis holds that any uncountable subset of the reals is equinumerous with the reals:

$$\mathfrak{c} = 2^{\aleph_0} = \aleph_1.$$

This is absolutely unsolvable, so far as we know.

To review that knowledge, let's assume that ZFC is consistent. We cannot prove its consistency from within ZFC, but let's hope it is, and assume it now. Then, we know that (1) ZFC + CH is consistent, and (2) ZFC + "CH *is false*" is consistent. (1) is due to Gödel. (2) is due to Cohen. The outcome is that CH is independent of the axioms of ZFC. It cannot be decided one way or the other as things stand. Let us take another look at our main model, L. If we use elementary embeddings, that model can be extended to a new model.

---

* [**128**]; [**129**]; [**130**]; [**91**]; [**56**]; [**61**].

2. $V =$ Ultimate L

Before I tell you about the new model, I remind you that the current state of affairs in unsatisfactory. At present, we do not have a single measure of the continuum.

Imagine there is a multiverse. Then CH holds true in some universes of set theory but is false in others. I have tried to be pluralistic about ways of knowing from the outset. But that does not extend to the formal structure of mathematical reality. There are many ways to define and understand the continuum. But there is only one continuum. But then, a rejection of the multiverse seems to be tied to a commitment, to CH being either true or false. "How am I to understand this? Did you not just say that this is absolutely unsolvable?" Yes. The current state of our knowledge is that it is absolutely unsolvable. If one is wedded to the idea that there is a solution, I seem to commit to a counterfactual, a state of expert knowledge that does not currently exist. This is your criticism. But not so fast. Many years ago Kurt Gödel thought about this very carefully, and had the insight that large cardinal axioms might be decisive. More recently others have come to agree with him, but for their own reasons. Their agreement on CH is based on new developments, and in particular on an advance in our technical mathematical knowledge.

Using elementary embeddings, one can extend models like L to incorporate large cardinals. And indeed L itself has been extended to a new model, Ultimate L. CH is true in Ultimate L. The current state of the art hypothesis is W. Hugh Woodin's hypothesis:

$$V = \text{Ultimate } L.$$

If that is true, then so is CH.

There are many ways to know about the continuum, and about sets. We have seen more than one theoretical account of the continuum. But there are not many continuums, one for each account. There are different ways to know the same thing.

3. Coda

We have seen persuasive accounts designed to represent our mathematical and scientific mentality taken somehow over the group of experts. I want to keep separate the real issue of our mentality in that precise sense from the real issue of our world. The universe is expanding, whether or not I recognize it. In this sense, it is no business of mine. Likewise, the mathematical continuum is composed of real numbers:

$$\mathfrak{c} = 2^{\aleph_0}.$$

That cannot solely be my business, for I am a finite being.

The real numbers consist of natural numbers, by definition. A real number is a subset of the natural numbers. These are our counter sets. The continuum is made up of subsets of the counter sets. These counter sets have nice sequential properties amongst them allowing for an intimate bond of continuity from one to the next. It is because of that continuity that insofar as we do conceive the continuum as stretches instead of points, this is not a mistaken or imperfect conception.

I have argued my account aligns better than others not only with our ordinary ways of talking about continua but also with some of our deepest intuitions about them. We have already talked about what continuous form progressions do, so I had better quit before I start repeating myself. I do not want to get any further behind than I am already.

APPENDIX A

Answers to Selected Exercises

(1) Assume $X \in \mathcal{P}Y$, Y includes X. Show that $\mathcal{P}X \subseteq \mathcal{P}Y$.
(2) Let $X = \{0, 1\}$ Using the definition of the ordered pair, deduce that $\langle 0, 1\rangle \subseteq \mathcal{P}X$
(3) Use *Aussonderong* to distinguish the "Wheat" pennies and the "Lincoln Memorial" pennies from the other coins in my pocket.
(4) Recall the counting set definitions from above. What is 3? Do 4 for extra credit.
(5) What is the rank of 2 in the hierarchy of sets? What about the rank of 3? Do you want to generalize?
(6) Show that the natural numbers are equinumerous with the positive integers.
(7) Show that $\mathbb{N} \sim \mathbb{Z}$.

1. Solutions

(1) Assume that $X \in \mathcal{P}Y$, i.e. that $X \subseteq Y$. Now, consider some random subset of X, say a. Thus by hypothesis, $a \in \mathcal{P}X$. Now I must show that $a \in \mathcal{P}Y$ by the definition of inclusion. So, the problem is saying that if $X \subseteq Y$, then subsets of X like a are subsets of Y. Well, by the assumption, all the members of X are elements in Y. But any element of a is an element of X because $a \subseteq X$. So, by the assumption, for every x,

$$x \in a \to x \in Y.$$

But then $a \subseteq Y$, by definition, so my proof is done.

(2) I must show that $\langle 0, 1\rangle \in \mathcal{P}\mathcal{P}X$. By the definition of the ordered pair, it suffices to show that $\{\{0\}, \{0, 1\}\} \in \mathcal{P}\mathcal{P}X$. By the definition of X and power set, it suffices to show that $\{\{0\}, \{0, 1\}\}$ $\subseteq \mathcal{P}\{0, 1\}$. But $\{0\} \in P\{0, 1\}$ and $\{0, 1\} \in P\{0, 1\}$, so I have completed the task.

(5) We have to be careful with the ordinal backbone of the universe V of sets, for it remains narrow while the universe widens dramatically in its ascent through the ranks. Consider that $2 = \{\emptyset, \{\emptyset\}\} = V_2$. V_2 has all the elements of 2, so $2 \in V_3$. $V_3 = \mathcal{P}V_2$,

so V_3 has four subsets: $\emptyset, \{\emptyset\}, \{\{\emptyset\}\}$, and $\{\emptyset, \{\emptyset\}\}$. So, $3 \subseteq V_3$, and $3 \in V_4$. But $3 \neq V_3$. I should generalize this result before finishing.

(6) I use the Hilbert Hotel. I have an infinite number of rooms numbered $1, 2, 3, \ldots$ and a guest in each room, named One, Two, Three, and so on. *The idea is to fit the natural numbers into the rooms. But where does 0 go?* How do we make space for this new resident, Zero? The system is to put the new resident in Room 1 (move Zero to Room 1), and move the current resident of Room 1 to Room 2 (move One to Room 2), and move the resident of Room 2 to Room 3, and move the resident of Room 3 to Room 4, and so on. Or I use the fact that the successor function allows me to map members from the first set 1-1 and onto the second set: $0 \mapsto 1$, $1 \mapsto 2$, and so on. That should take care of that.

(7) I construct the following series that integrates the integers into a single list, making them equinumerous with the set of natural numbers, as follows:

$$0, 1, -1, 2, -2, 3, -3, \ldots.$$

APPENDIX B

Diagonalization

In this appendix, we establish that there are fewer natural numbers than real numbers. We prove this inequality by a diagonalization argument. In our terms, this is a result about counter sets, see **3.4** above, and subsets of them. The subsets of counter sets are

$$\{\, r : r \subseteq \omega \,\} = 2^{\omega} = \mathbb{R}.$$

The cardinality of the subsets of the counter sets is

$$|\mathbb{R}| = 2^{\aleph_0}.$$

Every member of each r is a counter set.

The theorem is that

$$\aleph_0 < 2^{\aleph_0}.$$

There are fewer natural numbers than real numbers. The cardinality of the set of counter sets is less than that of the set of subsets of them.

This should surprise no one at this juncture. We have been talking about there being more sets of objects than there are objects since **2.1.2**. Here the generalization is to take "objects" as counter sets and "sets of objects" as "subsets of the counter sets."

Let us prove this theorem. When we pose the question of the number of points on the number line, due to **4.3.4** above, to answer this question it suffices to show that the cardinality of the set $\{\, r : 0 < r < 1 \,\}$ has a greater cardinality than that of the counter sets.

The proof is by reductio ad absurdum. For the sake of the reductio, assume that one can count all the real numbers between 0 and 1. So, list them, one on each line. To be listable is to be denumerable or countable. So, the assumption would be that there is a bijection between the natural numbers and the real numbers, that is, between the counter sets and subsets of them.

We build the list, treating each of the r between 0 and 1 as a decimal expansion, a series of natural numbers. Now the list of decimal expansions has been created. Maybe it looks like this:

11000...
12345...
33633...
57721...
78539...
and so on...

We can identify a real number between 0 and 1 that is not on the list.

The way to do this is as follows. First, design a rule that takes each numeral to its successor, except: $9 \mapsto 0$ and $6 \mapsto 9$. Call this rule R.

Then consider the number formed off the diagonal of the list. I construct the expansion of this number as follows. The first digit of this new number is made by taking the first digit of the first decimal expansion on the list and applying R to it. The second digit of this new number is the second digit of the second decimal expansion on the list once R. The third digit of this new number is the third digit of the third decimal expansion on the list once R is applied.

We can quickly see that this new decimal expansion '23930...' is nowhere on the original list. How?

Now, let me answer that for you. It cannot be the first expansion on the list because it differs from the first expansion in its first digit. It cannot be the second decimal expansion on the list because it differs from the second expansion in its second digit. And so on. It is not on the list. This is a contradiction.

That means that there are more subsets of the counter sets than can be counted. But the natural numbers are the counter sets. Therefore, there are fewer natural numbers than real numbers.

An implication is that infinity is not all one thing but is a matter of degrees of cardinality. Cantor established a series of infinities. There is a sequence of $\aleph$'s:

$$\aleph_0, \aleph_1, \aleph_2, \dots \aleph_\omega, \aleph_{\omega+1} \dots$$

Each is strictly less than the next. Which degree of infinity is the continuum? Diagonalization does not provide that answer.

APPENDIX C

Continuum Problem

How many subsets do the counter sets have? Well, how many real numbers are there on the number line? This is Cantor's Continuum Problem. It asks about $\mathfrak{c}$, the cardinality of the continuum

$$\mathfrak{c} = |\mathbb{R}|.$$

The Continuum Problem is Hilbert's first problem, by the way.*

Cantor faced the problem and proposed a solution, the Continuum Hypothesis. Cantor's Continuum Hypothesis holds that any uncountable subset of the reals is equinumerous with the reals:

$$\mathfrak{c} = 2^{\aleph_0} = \aleph_1.$$

This is absolutely unsolvable, so far as we know.

To review that knowledge, let's assume that ZFC is consistent. We cannot prove its consistency from within ZFC, but let's hope it is, and assume it now. Then, we know that (1) ZFC + CH is consistent, and (2) ZFC + "CH *is false*" is consistent. (1) is due to Gödel. (2) is due to Cohen. The outcome is that CH is independent of the axioms of ZFC. It cannot be decided one way or the other.

Using elementary embeddings, one can extend models like L to incorporate large cardinals. And indeed L itself has been extended to a new model, Ultimate L. CH is true in Ultimate L. The current state of the art hypothesis is W. Hugh Woodin's hypothesis:

$$V = \text{Ultimate } L.$$

If that is true, then the size of the continuum is $\aleph_1$.

* This appendix involves: [**128**]; [**129**]; [**130**]; [**91**]; [**56**]; [**61**].

Bibliography

[1] Anscombe, G.E.M. *An Introduction to Wittgenstein's* Tractatus. 3rd ed. Philadelphia: Univ. of Pennsylvania Press, 1971.

[2] Arthur, R.T.W. Book review of [**50**]. *Philosophia Mathematica (III)* 27 (2019): 148–152.

[3] Awodey, S., and A.W. Carus. How Carnap could have replied to Gödel. In Awodey and Klein (eds.), *Carnap Brought Home: The View from Jena.* La Salle: Open Court, 2004.

[4] Baldwin, J.T. Axiomatizing changing conceptions of the geometric continuum II: Archimedes-Descartes-Hilbert-Tarksi. *Philosophia Mathematica (III)* 27 (2019): 33–60.

[5] Bell, John L. *The Continuous and the Infinitesimal in Mathematics and Philosophy.* Milano: Polimetrica, 2006.

[6] ———. Types, sets and categories. In Gabbay, Kanamori, and Woods (eds.), *Sets and Extensions in the Twentieth Century.* Boston: North-Holland, 2012, pp. 633–687.

[7] Black, Max. *A Companion to Wittgenstein's "Tractatus."* Ithaca: Cornell UP, 1964.

[8] Boolos, George. The advantages of honest toil over theft. In *Logic, Logic, Logic*, ed. Jeffrey. Cambridge, Mass.: Harvard University Press, 1999.

[9] Buckley, B.L. *The Continuity Debate: Dedekind, Cantor, du Bois-Reymond, and Peirce on Continuity and Infinitesimals.* Boston: Docent, 2012.

[10] Cantor, Georg. *Grundlagen einer allgemeinen Mannigfaltigkeitslehre.* Leipzig: Teubner, 1883.

[11] Carnap, Rudolf. *Abriss der Logistik.* Vienna: Julius Springer, 1929.

[12] ———. *Der logische Aufbau der Welt.* Berlin-Schlachtensee: Weltkreis, 1928.

[13] ———. *Die Methode der logischen Analyse.* In *Actes du huitième Congrès international de philosophie, à Prague 2–7 Septembre 1934.* Prague: Orbis, 1936.

[14] ———. *Logische Syntax der Sprache,* trans. Smeaton. London: Kegan Paul, 1937. (German original, 1934.)

[15] ———. *Überwindung der Metaphysik durch logische Analyse der Sprache. Erkenntnis* 2 (1932): 219–241.

[16] Dedekind, Richard. *Stetigkeit und irrationale Zahlen.* Braunschweig: F. Vieweg, 1872. Trans. in [**34**].

[17] ———. *Was sind und was sollen die Zahlen?* 1888. Trans. in [**34**].

[18] De Risi, V., ed. *Mathematizing Space: The Objects of Geometry from Antiquity to the Early Modern Age.* New York: Birkhäuser, 2015.

[19] Dewey, John. *Human Nature and Conduct.* New York: Random House, 1957.

[20] Dreben, Burton. Quine. In Barrett and Gibson (eds.), *Perspectives on Quine.* Oxford: Blackwell, 1990.

[21] Ebbs, Gary. *Carnap, Quine, and Putnam on Methods of Inquiry.* Cambridge: Cambridge UP, 2017.
[22] Elgin, Catherine Z. *Considered Judgment.* Princeton: Princeton UP, 1996.
[23] ______. Construction and cognition. *Theoria* 65 (2009): 135–146.
[24] ______. Emotions and understanding. In Brun and Doguoglu (eds.), *Epistemology and Emotions.* New York: Routledge, 2008, ch. 1.
[25] ______. L'exemplification et la danse. In Beauquel and Pouivet (eds.), *Philosophie de la Danse.* Rennes: Presses Universitaires de Rennes, 2010, pp. 81–98.
[26] ______. Fiction as thought experiment. *Perspectives on Science* 22 (2014): 221–241.
[27] ______. The laboratory of the mind. In Gibson, Huemer, and Pocci (eds.), *A Sense of the World: Essays on Fiction, Narrative, and Knowledge.* New York: Routledge, 2007, pp. 43–54.
[28] ______. Nominalism, realism and objectivity. *Synthese* 196 (2019): 519–534.
[29] ______. Reference and Meaning: A Tractarian Analysis of Incommensurable Representational Systems. Ph.D. dissertation, Brandeis University, Department of Philosophy, October, 1974.
[30] ______. Scheffler's symbols. *Synthese* 94 (1993): 3–12.
[31] ______. *True Enough.* Cambridge, Mass.: MIT Press, 2017.
[32] ______. *With Reference to Reference.* Cambridge, Mass.: Hackett, 1983.
[33] Enderton, Herbert B. *Elements of Set Theory.* New York: Academic, 1977.
[34] Ewald, W.B., ed. *From Kant to Hilbert: A Source Book in the Foundations of Mathematics.* Vol. II. Oxford: Clarendon, 1996.
[35] Feferman, Solomon. Conceptions of the continuum. *Intellectica* 51 (2009): 169–189.
[36] ______. Hilbert's program relativized: proof-theoretical and foundational reductions. *The Journal of Symbolic Logic* 53 (1988): 364–384.
[37] ______. *In the Light of Logic.* New York: Oxford University Press, 1998.
[38] Frege, Gottlob. *Begriffsschrift, eine der arithmetischen nachgebildete Formelsprache des reinen Denkens.* Halle: 1879. Trans. van Heijenoort in [**117**].
[39] ______. *Die Grundlagen der Arithmetik. Eine logisch-mathematische Untersuchung über den Begriff der Zahl.* Breslau: Koebner, 1884.
[40] ______. *Grundgesetze der Arithmetik, begriffsschriftlich abgeleitet.* Vol. 1. Jena: Pohle, 1893. Vol. 2 (Jena: Pohle, 1903).
[41] Gandy, R.O. Bertrand Russell, as mathematician. *Bulletin of the London Mathematical Society* 5 (1973): 342–348.
[42] Geach, Peter. Wittgenstein's operator N. *Analysis* 41 (1983): 573–589.
[43] Gödel, Kurt. *Collected Works.* Vol. I, ed. Feferman etal. New York: Oxford UP, 1986.
[44] ______. *Collected Works.* Vol. II, ed. Feferman etal. New York: Oxford UP, 1995.
[45] ______. Consistency proof for the generalized Continuum Hypothesis. 1995. Reprinted in [**44**].
[46] ______. On formally undecidable propositions of *Principia Mathematica* and related systems I. 1931. Reprinted in [**43**], pp. 144–195.
[47] ______. Russell's mathematical logic. 1944. Reprinted in [**44**].

[48] ———. What is Cantor's continuum problem? 1947. Rev. ed. (1964). Reprinted in [**44**].

[49] Goldfarb, Warren. Russell's reasons for ramification. In Savage and Anderson (eds.), *Rereading Russell: Essays on Bertrand Russell's Metaphysics and Epistemology*. Minneapolis: University of Minnesota Press, 1989.

[50] Hellman, G., and S. Shapiro. *Varieties of Continua. From Regions to Points and Back.* Oxford: Oxford UP, 2018.

[51] Hintikka, Jaakko, ed. *From Dedekind to Gödel: Essays on the Development of the Foundations of Mathematics.* Dordrecht: Kluwer, 1995.

[52] ———. Identity, variables, and impredicative definitions. *The Journal of Symbolic Logic* 21 (1956): 225–245.

[53] Husserl, E. *Ideas Pertaining to a Pure Phenomenology and to a Phenomenological Philosophy,* trans. Kersten. The Hague: Nijhoff, 1982.

[54] ———. *Logical Investigations,* trans. Findlay. London: Routledge and Kegan Paul, 1970.

[55] Hylton, Peter, and Gary Kemp. Willard Van Orman Quine. In Zalta and Nodelman (eds.), *Stanford Encyclopedia of Philosophy,* 2025.

[56] Kanamori, Akihiro. *The Higher Infinite. Large Cardinals in Set Theory from Their Beginnings.* New York: Springer, 1994. 2nd corr. ed. (2005).

[57] ———. Putnam's constructivization argument. In Hellman and Cook (eds.), *Putnam on Mathematics and Logic.* Berlin: Springer, 2018.

[58] Keyser, C.J. 1903. Concerning the axiom of infinity and mathematical induction. *Bulletin of the American Mathematical Society* 9 (1903): 424–434.

[59] ———. On the axiom of infinity. *Hibbert Journal* 2 (1904): 532–552.

[60] Koellner, Peter. The continuum hypothesis. In Zalta and Nodelman (eds.), *The Stanford Encyclopedia of Philosophy.* 2023.

[61] ———. Large cardinals and determinacy. In Zalta (ed.), *The Stanford Encyclopedia of Philosophy.* 2014.

[62] Kripke, Saul. Russell's notion of scope. *Mind* 114 (2005): 1005–1037.

[63] Kuratowski, C. Sur la notion de l'ordre dans la Théorie des Ensembles. *Fundamenta Mathematicae* 2 (1921): 161–171.

[64] Leibniz, G.W. *The Labyrinth of the Continuum: Writings on the Continuum Problem, 1672–1686,* ed. Arthur. Yale: Yale UP, 2001.

[65] Link, Montgomery. A background condition for Leibniz's philosophy. In Pisano, etal. (eds.), *The Dialogue between Sciences, Philosophy and Engineering. Homage to Gottfried W Leibniz 1646 – 1716.* London: College Publications, 2017, pp. 227–254.

[66] ———. The dispute between two accounts of the continuum. *The Journal of Philosophy* 119 (2022): 425–443.

[67] Linsky, B. Russell's theory of definite descriptions and the idea of logical construction. In Beaney (ed.), *The Oxford Handbook of the History of Analytic Philosophy.* Oxford: Oxford UP, 2013.

[68] Marion, Mathieu. *Ludwig Wittgenstein. Introduction au Tractatus logico-philosophicus.* Paris: Presses Universitaires de France, 2004.

[69] ———. Operations and numbers in the Tractatus. In Oliveri (ed.), *From the Tractatus to the Tractatus and Other Essays.* Wittgenstein Studien, vol. 2. New York: Peter Lang, 2001, pp. 105–123.

[70] ———. Wittgenstein and Ramsey on identity. In [**51**], 343–373. 1995,

[71] McGuinness, Brian. The so-called realism of Wittgenstein's Tractatus. In Block (ed.), *Perspectives on the Philosophy of Wittgenstein.* Cambridge, Mass.: MIT Press; 1981.

[72] Miller, F.D. Aristotle against the Atomists. In Kretzmann (ed.), *Infinity and Continuity in Ancient and Medieval Thought*. Ithaca: Cornell UP, 1982.

[73] Monk, Ray. *Bertrand Russell: The Spirit of Solitude, 1872-1921*. New York: Free Press, 1996.

[74] Moore, Gregory H. The origins of Russell's paradox: Russell, Couturat, and the antinomy of infinite number. In [**51**]. 1995.

[75] Parsons, Charles. Developing arithmetic in set theory without infinity: some historical remarks. *History and Philosophy of Logic* 8 (1987): 201–213.

[76] ———. *From Kant to Husserl. Selected Essays*. Cambridge, Mass.: Harvard UP, 2012.

[77] ———. *Philosophy of Mathematics in the Twentieth Century. Selected Essays*. Cambridge, Mass.: Harvard UP 2014.

[78] Putnam, Hilary. The analytic and the synthetic. 1962. Reprinted in *Philosophical Papers*, vol. 2 (Cambridge: Cambridge UP, 1975).

[79] ———. Models and reality. *The Journal of Symbolic Logic* 45 (1980): 464–482.

[80] ———. A note on constructible sets of integers. *Notre Dame Journal of Formal Logic* 4 (1963): 229–308.

[81] ———. What is mathematical truth? *Historia Mathematica* 2 (1975): 529–545.

[82] Quine, Willard van Orman. *From a Logical Point of View*. Cambridge, Mass.: Harvard UP, 1953. 2nd ed. rev. (1980).

[83] ———. *From Stimulus to Science*. Cambridge, Mass.: Harvard UP, 1995.

[84] ———. New foundations for mathematical logic. *The American Mathematical Monthly* 44 (1937): 70–80. Reprinted ([**82**], pp. 80–101).

[85] ———. *The Roots of Reference*. La Salle: Open Court, 1974.

[86] ———. *Set Theory and Its Logic*. Rev. ed. Cambridge, Mass.: Belknap, 1969. 1st ed. (1963).

[87] ———. *The Ways of Paradox and Other Essays*. Rev. ed. Cambridge, Mass.: Harvard UP, 1976.

[88] ———. *Word and Object*. Cambridge, Mass.: MIT Press, 1960.

[89] Rawls, John. *A Theory of Justice* Cambridge, Mass.: Belknap, 1971.

[90] Reeder, P. Labyrinth of continua. *Philosophical Mathematica (III)* 26 (2017): 1–39.

[91] Rittberg, Colin J. How Woodin changed his mind: new thoughts on the Continuum Hypothesis. *Archive for History of Exact Sciences* 69 (2015): 125–151.

[92] Russell, Bertrand. *The Analysis of Matter*. New York: Harcourt, Brace, 1927.

[93] ———. The axiom of infinity. In [**96**] .

[94] ———*A Critical Exposition of the Philosophy of Leibniz: With an Appendix of Leading Passages*. Cambridge: at the University Press, 1900.

[95] ———*An Essay on the Foundations of Geometry*. Cambridge: Cambridge UP 1897. Reprinted (New York: Routledge, 1996).

[96] ———. *Essays in Analysis,* ed. Lackey. New York: George Brazier, 1973.

[97] ———. *Introduction to Mathematical Philosophy*. London: Unwin, 1919. Reprinted (New York: Routledge, 1993.)

[98] ———. Is mathematics purely linguistic? 1951. Reprinted in [**96**].

[99] ———. Logical atomism. 1924. Reprinted in [**100**].

[100] ———. *Logic and Knowledge. Essays, 1901–1950,* ed. Marsh. London: Unwin, 1956.
[101] ———. The logic of relations with some applications to the theory of series. 1901. Reprinted in [**100**] and [**112**].
[102] ———. Mathematical logic as based on the theory of types. 1908. Reprinted in [**100**] and [**117**].
[103] ———. *My Philosophical Development.* New York: Simon and Schuster, 1959.
[104] ———. On denoting. 1905. Reprinted in [**100**].
[105] ———. On matter. 1912. Reprinted (*Logical and Philosophical Papers. 1909–13,* ed. Slater. Collected Papers 6. New York: Routledge, pp. 77–95).
[106] ———. On some difficulties in the theory of transfinite numbers and order types. *Proceedings of the London Mathematical Society,* s. 2, 4 (1907): 29–53.
[107] ———. *Our Knowledge of the External World as a Field for Scientific Method in Philosophy.* New York: Open Court, 1914.
[108] ———. *Philosophical Papers. 1896–99,* ed. Griffin and Lewis. Collected Papers 2. New York: Routledge, 1990.
[109] ———. *The Philosophy of Logical Atomism and Other Essays. 1914–19,* ed. Slater. Collected Papers 8. Boston: George Allen & Unwin, 1986.
[110] ———. *Portraits from Memory and Other Essays.* New York: Simon and Schuster, 1956.
[111] ———. *Principles of Mathematics.* 2nd ed. New York: W.W. Norton, 1938. 1st ed. (1903).
[112] ———. *Toward the 'Principles of Mathematics,'* ed. Moore. Collected Papers 3. New York: Routledge, 1993.
[113] Sheffer, H.M. Review of [**123**]. *Isis* 8 (1926): 226 – 231.
[114] Stečkin, S.B. Theory of functions of a real variable. Chapter XV in Aleksandrov, etal. (eds.), *Mathematics. Its Content, Methods and Meaning. Three Volumes Bound as One,* trans. S.H. Gould, etal. Garden City, NY: Dover, 2024.
[115] Suppes, Patrick. *Introduction to Logic.* Princeton: D. van Nostrand, 1957.
[116] Urquhart, A. The theory of types. In Griffin (ed.), *The Cambridge Companion to Bertrand Russell.* Cambridge: Cambridge UP, 2003.
[117] van Heijenoort, Jean, ed. *From Frege to Gödel: A Source Book in Mathematical Logic, 1879–1931.* Cambridge, Mass.: Harvard UP, 1967.
[118] Webb, Judson. Tracking contradictions in geometry: the idea of a model from Kant to Hilbert. In [**51**]. 1995.
[119] Weyl, Hermann. *Das Kontinuum. Kritische Untersuchungen über die Grundlagen der Analysis.* Leipzig: Veit, 1918.
[120] White, M.J. *The Continuous and the Discrete. Ancient Philosophical Theories from a Contemporary Perspective.* Oxford: Oxford UP, 1992.
[121] Whitehead, Alfred North, and Bertrand Russell. *Principia Mathematica.* Vol. 1. Cambridge: at the Univ. Press, 1910
[122] ———. *Principia Mathematica.* Vol. 2. Cambridge: at the Univ. Press, 1912
[123] ———. 2nd. ed. of [**121**]. Cambridge: Cambridge UP, 1925
[124] Wiener, Norbert. A simplification of the logic of relations. 1914. Reprinted in [**117**] .

[125] Wittgenstein, Ludwig. *Cambridge Letters. Correspondence with Russell, Keynes, Moore, Ramsey, and Sraffa*, ed. McGuinness and von Wright. Oxford: Blackwell, 1995.

[126] ———. *Philosophical Grammar,* ed. Rhees, trans. Kenny. Los Angeles: Univ. of California Press, 1974.

[127] ———. *Tractatus Logico-Philosophicus,* trans. Ogden. London: Kegan Paul, 1922.

[128] Woodin, W. Hugh. The continuum hypothesis. Part I. *Notices of the AMS* 48 (2001): 567–576.

[129] ———. Suitable extender models I. *Journal of Mathematical Logic* 10 (2010): 101–339.

[130] ———. Suitable extender models II: beyond ω-huge. *Journal of Mathematical Logic* 11 (2011): 115–436.

[131] Zermelo, Ernst. Sur les ensembles finis et le principe de l'induction complete. *Acta Mathematica* 32 (1909): 185–193. 1909.

Index – Names

Index – Subjects

This book on sets and the continuum has three parts. The first part lays the set-theoretical and logical groundwork; the second presents the analysis of the continuum itself; and the third reflects on the key ontological and epistemological debates.

The first part offers a gentle introduction to sets, a basis for the discussions that follow. But the reader should be forewarned that there is more going on. From the empty set to the innocent abstraction principle, the text inculcates the principles of set theory.

The second part concerns the analysis of the continuum. The continuum is divisible infinitely. But there is a great dispute over whether the continuum is divisible all the way down. It might instead consist of indivisible points that are bonded together into a continuity.

The third part then turns to three related debates. The first debate concerns set-theoretic ontology. Link argues that set theory is not superfluous. The second debate is between realism and nominalism. The third debate concerns logical construction and the nature of scientific reduction. The overall argument lies within this third debate.

The overall argument is highly controversial. There is not one way or one best way to know the continuum. Any reduction qualifies, once it meets the formal and mathematical requirements, as a legitimate way of knowing the continuum. There are many ways to know it; however, the continuum is explained in set theory. It is real, if the sets are. Indeed, that is what makes the book controversial.

Montgomery Link is Associate Professor, Suffolk University. He lives with his wife and four children in Bullfinch Triangle in Boston and walks to work on Beacon Hill.

www.ingramcontent.com/pod-product-compliance
Ingram Content Group UK Ltd.
Pitfield, Milton Keynes, MK11 3LW, UK
UKHW021239300926
15471UKWH00013B/764